Dedication

In memory of our parents. This book is dedicated to those
we teach and those who taught us.

Acknowledgments

We are grateful to the team at Wolters Kluwer. We thank Crystal Taylor whose support has been invaluable throughout the course of the first and second editions of this book. We also extend special thanks to Matt Chansky who expertly and patiently transformed our sketches into the works of art that appear as figures throughout this book.

We value the support of our colleagues who read and critiqued early versions of the chapters, who served as peer reviewers, and to those who have adopted this title for their courses. We hope that this second edition will be a helpful resource to students of the health professions.

In Memoriam

Richard A. Harvey, PhD

1936–2017

Cocreator and series editor of the Lippincott Illustrated Reviews series,
in collaboration with Pamela C. Champe, PhD (1945–2008).

Illustrator and coauthor of the first books in the series:
Biochemistry, Pharmacology, and *Microbiology and Immunology.*

Contents

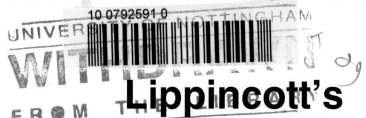

Lippincott's Illustrated Reviews: Cell and Molecular Biology

Second Edition

Nalini Chandar, PhD

Professor of Biochemistry

Midwestern University

Downers Grove, Illinois

Susan Viselli, PhD

Professor of Biochemistry

Midwestern University

Downers Grove, Illinois

Wolters Kluwer

Philadelphia · Baltimore · New York · London
Buenos Aires · Hong Kong · Sydney · Tokyo

Acquisitions Editor: Crystal Taylor
Product Development Editor: Andrea Vosburgh
Marketing Manager: Michael McMahon
Production Project Manager: Marian Bellus
Editorial Coordinator: Dave Murphy
Design Coordinator: Joan Wendt
Manufacturing Coordinator: Margie Orzech
Prepress Vendor: SPi Global

Second Edition

Library of Congress Cataloging-in-Publication Data
Names: Chandar, Nalini, author. | Viselli, Susan, author.
Title: Cell and molecular biology / Nalini Chandar, Susan Viselli.
Other titles: Lippincott's illustrated reviews.
Description: Second edition. | Philadelphia : Wolters Kluwer, [2019] | Series: Lippincott's illustrated reviews | Includes index.
Identifiers: LCCN 2017054851 | ISBN 9781496348500
Subjects: | MESH: Cells | Biological Transport—physiology | Cell Physiological Phenomena | Genetic Phenomena | Examination Questions | Outlines
Classification: LCC QH581.2 | NLM QU 18.2 | DDC 611/.0181—dc23 LC record available at https://lccn.loc.gov/2017054851

CCS0118

UNIT I
Cell and Tissue Structure and Organization

Unity of plan everywhere lies hidden under the mask of diversity of structure—the complex is everywhere evolved out of the simple.
—Thomas Henry Huxley
A Lobster; or, the Study of Zoology (1861). In: *Collected Essays*, Vol. 8. 1894: 205–206.

The most basic and simple form of human life is the cell. Complex organisms, such as humans, are collections of these individual cells that have grown and differentiated to generate the organism itself. Each cell in an adult developed in a purposeful way from a precursor cell along a specific lineage to become organized structurally and according to its function. Be it a liver cell, a blood cell, a bone cell, or a muscle cell, it was derived from a stem cell generated soon after the organism's conception. Hidden within the tiny stem cell is the vast capacity to develop into an array of diverse cell types that differentiate into efficient individual units that survive only within the context of the whole person.

Our discussion of cell and molecular biology therefore begins with an examination of stem cells, from which all other cells are generated. As we move deeper into this unit, we explore structural components of tissues including the extracellular matrix, which is produced by cells but resides outside the boundary of cell membranes. In considering cell structure, cell membranes serve as our starting point. As the outer limit of the cell, the plasma membrane protects the cell's interior from the environment. Yet, it is also a dynamic structure that enables interactions with the environment and facilitates the cell's function. Within the confines of the plasma membrane, we find cytoskeletal proteins that not only organize the cytoplasm and provide a structural framework for the cell but also function in the intracellular movement of chromatin and organelles. The organelles are specialized centers within each cell that carry out functional processes, including energy extraction in the mitochondria, macromolecule digestion in the lysosomes, and DNA and RNA synthesis in the nucleus. While each organelle is a complex machine in its own right and has its own unique role within the cell, organelles are linked physically by the cytoskeleton and cooperate in completing their tasks with a unified purpose.

1 Stem Cells and Their Differentiation

I. OVERVIEW

All cells within an organism are derived from **precursor cells**. Precursor cells divide along specific paths in order to produce cells that are differentiated to perform specialized tasks within tissues and organs. Cells with the capacity to give rise to the whole organism are referred to as **stem cells**. Stem cells remain undifferentiated and are characterized by the ability to self-renew. They also generate many daughter cells committed to differentiate (change) into a diverse range of specialized cell types. A daughter cell with the ability to differentiate into a wide variety of cell types is **pluripotent**.

The human body is made up of about 200 different types of cells. The human genome is the same in all cell types, meaning that within an individual person, all cells have exactly the same DNA sequences and genes. Stem cells represent all the different ways that human genes can be expressed as proteins. In order for different regions of the genome to be expressed in different cell types, the genome has to be reversibly modified. In fact, the organization of **chromatin** (a complex of specific proteins and DNA) varies between cell types and is made possible by reversible covalent modification of the proteins that are associated with DNA and sometimes the DNA itself (see also Chapter 6). These modifications are important to expose regions of the DNA to the proteins and enzymes that are required for **transcription** (DNA → RNA), allowing for variation in gene expression as proteins (see also Chapter 8).

Different cell types arise from precursor cells that proliferate (divide) and eventually differentiate into cells with unique structures, functions, and chemical composition. From specific proteins produced in the cells, a special type of cell division, and the microenvironment of the precursor cell, a progeny of cells is produced. This progeny can both sustain the organism and carry out specific functions in the body.

II. STEM CELLS

Stem cells exist in both early embryos and adult tissues (Figure 1.1). The stem cells present in early embryos have an extensive capacity to differentiate into all cell types of the organism. Populations of stem cells also exist in adults and can differentiate into various cells within a lineage but not outside that lineage. For example, a hematopoietic stem cell (HSC) can differentiate into various types of blood cells but not into a hepatocyte (liver cell). However, researchers may have found new properties of adult

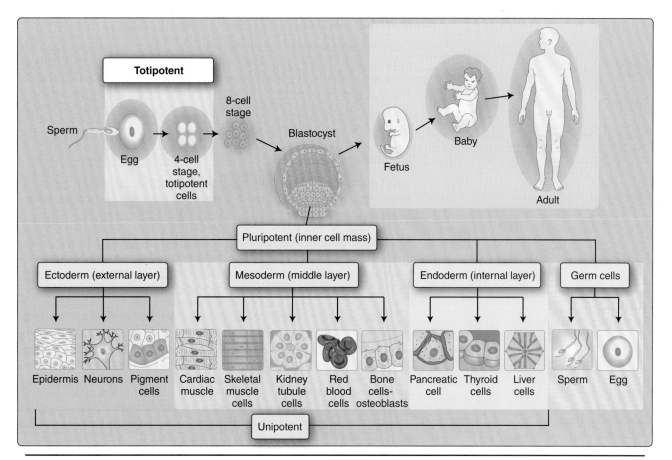

Figure 1.1
Embryonic and adult stem cells.

stem cells, which under certain conditions allow differentiation into more cell types than previously recognized.

- **Totipotency** is the potential of a single cell to develop into a total organism (e.g., a fertilized egg and the four-cell stage).
- **Pluripotency** is a cell's capacity to give rise to all cell types in the body but not to the supporting structures, such as the placenta, amnion, and chorion, all of which are needed for the development of an organism.
- **Multipotency** is a cell's capacity to give rise to a small number of different cell types.
- **Unipotency** is a cell's capacity to give rise to only one cell type.

A. Pluripotent stem cells

The most primitive, undifferentiated cells in an embryo are **embryonic stem cells** (ESCs). These cells are derived from the inner cell mass of preimplantation embryos (Figure 1.1) and can be maintained in the pluripotent state in vitro. They can give rise to all the three germ layers (ectoderm, mesoderm, and endoderm) with the ability to differentiate into a number of cell types. This ability to differentiate into multiple cell types is called **plasticity** (Figure 1.2).

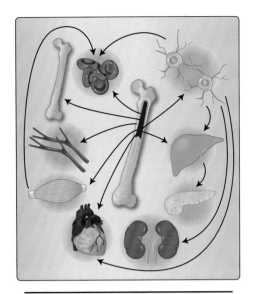

Figure 1.2
Adult stem cell plasticity.

B. Unipotent stem cells

Cells that reside within adult tissues and retain the ability to generate cells for the tissue type to which they belong are unipotent. Under normal situations, unipotent stem cells give rise to only one cell type. For example, a myoblast (muscle cell precursor) can give rise to a myocyte (muscle cell), and a hepatoblast (liver cell precursor) can give rise to a hepatocyte (liver cell).

C. Multipotent stem cells

Adult multipotent stem cells have been identified for a number of different tissue types such as brain, bone marrow, peripheral blood, blood vessels, skeletal muscle, skin, and liver (Figure 1.2).

1. **Hematopoietic stem cells:** This group of stem cells gives rise to all the types of blood cells, including red blood cells, B lymphocytes, T lymphocytes, natural killer cells, neutrophils, basophils, eosinophils, monocytes, macrophages, and platelets.

2. **Mesenchymal stem cells:** Also called **bone marrow stromal cells**, mesenchymal cells give rise to a variety of cell types, including osteoblasts (bone cells), chondrocytes (cartilage cells), adipocytes (fat cells), and other kinds of connective tissue.

3. **Skin stem cells:** These stem cells are found in the basal layer of the epidermis and also at the base of hair follicles. Epidermal stem cells give rise to keratinocytes, while the follicular stem cells give rise to both hair follicles and the epidermis.

4. **Neural stem cells:** Stem cells in the brain can differentiate into its three major cell types: nerve cells (neurons) and two types of non-neuronal cells—astrocytes and oligodendrocytes.

5. **Epithelial stem cells:** Located in the lining of the digestive tract, epithelial stem cells are found in deep crypts and give rise to several cell types, including absorptive cells, goblet cells, Paneth cells, and enteroendocrine cells.

III. STEM CELL COMMITMENT

Most stem cells first become intermediate progenitor cells (also known as **transit amplifying cells**), which then produce a differentiated population of cells. A good example of this stepwise process is illustrated by HSCs. HSCs are multipotent but undergo commitment into a particular pathway in a stepwise process. In the first step, they give rise to two different **progenitors**. The committed progenitors then undergo many rounds of cell division to produce a population of cells of a given specialized type. In this particular case, the HSCs give rise to one type that is capable of generating lymphoid cells and another type that results in the generation of myeloid cells (Figure 1.3).

The different steps of commitment happen due to changes in gene expression. Genes for a particular pathway are switched on, while access to other developmental pathways is shut off by specific proteins that bind DNA and act as **transcription factors** (see Chapter 10). These proteins are able to both activate the genes necessary for a particular pathway and shut down the expression of genes necessary for development along a different pathway.

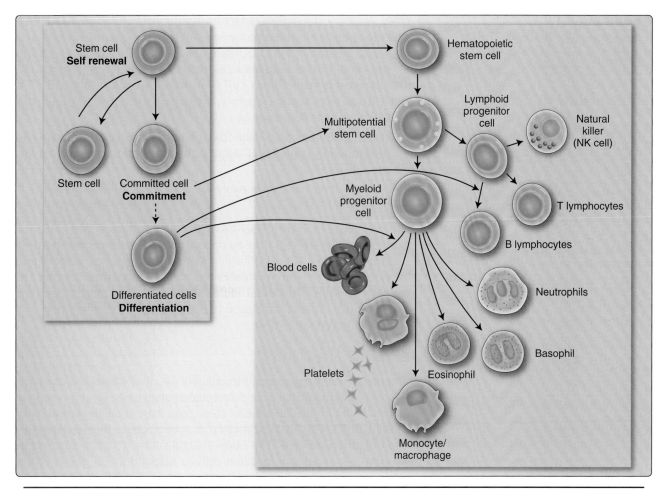

Figure 1.3
Stem cells commit to different pathways in a stepwise process.

IV. STEM CELL PLURIPOTENCY

In order to maintain a stable population of stem cells that are capable of self-renewal, mechanisms that prevent differentiation and promote proliferation must be transmitted to their daughter cells. While specific mechanisms by which stem cells retain their pluripotency remain largely unknown, studies with mouse ESCs suggest the importance of a self-organizing network of transcription factors that prevent differentiation and promote proliferation of stem cells (see below). Another alteration that facilitates this process is **epigenetic modification** (change that affects the ability of DNA to be transcribed to RNA without directly modifying the DNA sequence) of DNA, histones, or the chromatin structure in such a fashion that it changes the accessibility to binding by transcription factors.

A. Transcription factors

Among the various forms of regulation that orchestrate the cellular pluripotency program, transcriptional regulation appears to be the dominant form. In stem cells, transcription factors function through

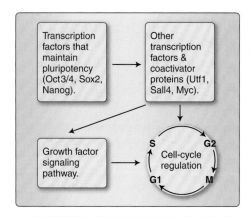

Figure 1.4
Proteins involved in maintaining stem cell pluripotency.

other ancillary transcription factors and coactivator proteins. These activate specific signaling pathways that lead to both cell survival and entry into the **cell cycle** to facilitate cell division. Several transcription factors act as master regulators to maintain pluripotency (Figure 1.4). The **core embryonic stem cell pluripotency factors** appear to be Oct4, Sox2, and Nanog. Together, these pluripotency factors serve to establish a pluripotent state by:

• Activating expression of other pluripotency-associated factors while simultaneously repressing target gene expression required for lineage differentiation.

• Activating their own gene expression while sustaining the expression of one another.

Recent advances have deciphered signaling pathways that influence pluripotency. For example, the mitogen-activated protein kinase (MAPK) pathway (Chapter 18) negatively influences pluripotency, while signal transducer and activator of transcription (STAT) signaling (Chapter 18) complements the pluripotent state. MicroRNAs (Chapter 8) are now understood to repress translation of selected mRNAs in stem cells and differentiating daughter cells.

B. Epigenetic mechanisms

When stem cells are induced to differentiate, their nuclei are strikingly different from the nuclei of undifferentiated stem cells. The chromatin is more relaxed in undifferentiated stem cells than in differentiated cells, allowing for the low-level expression of several genes that is characteristic of pluripotent cells. The open structure allows for the rapid regulation that is necessary for stem cells to respond to the needs of the organism. The core pluripotency factors (Oct4, Sox2, and Nanog) can modulate the chromatin states by regulating **DNA methyltransferases, Polycomb group proteins** and other chromatin remodeling factors. Thus, small changes in the levels of core factors such as Oct4 or Sox2 can determine if pluripotency is maintained or differentiation is triggered.

V. STEM CELL RENEWAL

Development requires cells to commit to different fates. However, in the case of stem cells, there must be a mechanism in place to maintain their populations while also generating differentiated populations. This mechanism is called **asymmetric cell division**.

Asymmetric cell division occurs when two daughter cells are produced that differ in fate. A stem cell has the capacity to produce one cell similar to itself (i.e., to continue being a stem cell) and to also generate another daughter cell that can proceed on a different pathway and differentiate into a particular type of cell (Figure 1.5).

Several mechanisms exist within stem cells that determine whether asymmetric cell division will take place. One such mechanism is **cell polarity**. Polarity is a stable feature in early embryos but may be a transient feature in tissue stem cells. External signals transmitted by receptors with seven transmembrane domains are involved in this process (see Chapter 17).

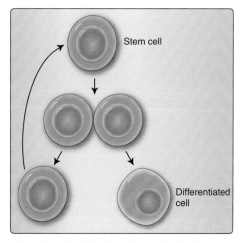

Figure 1.5
Asymmetric cell division.

VI. STEM CELL NICHE

If a population of stem cells has to be maintained as stem cells and not differentiate into any particular cell type, mechanisms must exist to guarantee their continued presence. The microenvironment that controls the self-renewal and maintenance of stem cells is termed the "**stem cell niche**." The niche saves stem cells from depletion while protecting the host from overproduction of stem cells. It performs as a basic tissue unit integrating signals that allow for a balanced response of stem cells to the needs of the organism. Progress has been made in the understanding and identification of niches for various types of tissue stem cells and in elucidating the role of the niche in regulating the asymmetric stem cell division. The choice of fate of a stem cell is determined by both extrinsic signaling and intrinsic mechanisms (see also Chapters 17 and 18).

A. Extrinsic signaling

While the cues that control proliferation and renewal of stem cells are not yet well defined, extracellular matrix interactions are known to play an important role (see also Chapter 2 for information regarding the extracellular matrix). Studies have highlighted the requirement for E-cadherin– and β-catenin–containing junctions between stem cells and cells that support their **stemness** or ability to self-renew and also maintain ability to differentiate (see also Chapter 2 for information about cell adhesion molecules and cell junctions). This is best understood with HSCs that associate with certain osteoblasts in the bone microenvironment that supports them. Signaling pathways that are engaged through adherens junctions have been found to be important in HSC renewal and proliferation. Cells outside of the direct contact with osteoblasts will differentiate, while the ones associated with this subpopulation of osteoblasts will remain as stem cells (Figure 1.6).

B. Intrinsic mechanisms

The mechanism of asymmetric cell division in mammalian cells is not completely understood and different mechanisms may result in asymmetric cell division. One such mechanism may rely on specific proteins to subdivide the cell into different domains before **mitosis** (nuclear division) so that an axis of polarity is generated. Segregation of certain cell-fate–determining molecules into only one of the two daughter cells will help maintain one as a stem cell and allow the other to differentiate (Figure 1.7).

VII. STEM CELL TECHNOLOGY

A. Regenerative medicine

The use of pluripotent stem cells offers the possibility of a renewable source of replacement cells and tissues to treat a number of diseases and conditions, including:

- Type 1 diabetes mellitus, where the β cells of the pancreas are destroyed.

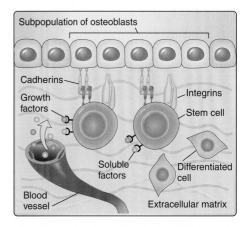

Figure 1.6
Extrinsic pathways maintaining HSC stem cell niche.

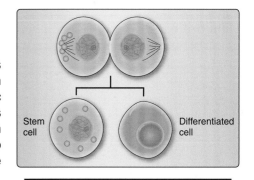

Figure 1.7
Differential segregation of cellular constituents during asymmetric cell division.

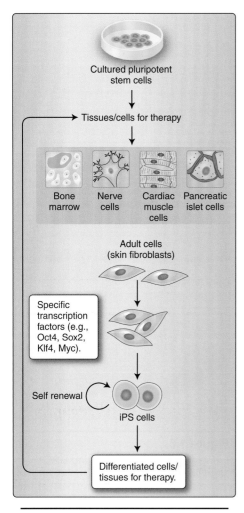

Figure 1.8
Stem cell–based therapy.

- Parkinson disease, where brain cells secreting dopamine are destroyed.
- Alzheimer disease, where neurons are lost due to accumulation of protein plaques in the brain.
- Stroke, where a blood clot causes oxygen deprivation and loss of brain tissue.
- Spinal cord injuries, leading to paralysis of skeletal muscles.
- Other conditions such as burns, heart disease, osteoarthritis, and rheumatoid arthritis, where lost cells can be replaced using stem cells.

B. Induced pluripotent stem cells and reprogramming

It has been possible to turn adult somatic cells into a pluripotent state by introducing extra copies of core genes that control pluripotency and is referred to as **cellular reprogramming**. Reprogramming has opened up the exciting prospect of being closer to treating disease using the patient's own somatic cells. In these cases, a cocktail of four genes, some of which constitute the core transcription factors in ESC maintenance (Oct4, Sox2, Klf4, and Myc) (OSKM cocktail), is sufficient to induce reprogramming of mouse skin fibroblasts into pluripotent cells called induced pluripotent stem cells (iPSCs). Reprogramming has also been attempted with several human cell types demonstrating the simplicity and reproducibility of this methodology. While there are several barriers to the reprogramming process, the potential of iPS cells in the treatment of the above-mentioned disorders is enormous (Figure 1.8). Several human clinical trials are ongoing to evaluate human ESC and iPSC-derived lines for use in spinal cord injury, macular degeneration, type 1 diabetes, Parkinson disease, and heart failure.

Clinical Application 1.1: Cord Blood and Tooth Banking

Cord blood, which was previously routinely discarded after childbirth, is now recognized as an important source of HSCs. It can now be been banked as a valuable tool for patients who may develop blood disorders. Thousands of cord blood transplantations have been performed to date for hematological malignancies and disorders such as leukemia, thalassemia, and sickle cell disease.

In the case of teeth, the presence of different types of mesenchymal stem cell populations in them has been described and labeled according to the harvest site. Dental pulp stem cells (DPSCs) from the mandibular third molar (wisdom tooth), stem cells from human exfoliated deciduous teeth (SHED) (baby teeth), and periodontal ligament stem cells (PDLSC) are some examples. These cells are generically referred to as dental stem cells (DSCs) and despite their small size are a source of abundant cells due to their highly proliferative nature. The ease of obtaining them by routine tooth extraction combined with the potential of these mesenchymal stem cells to differentiate into osteoblasts, chondroblasts, adipocytes, and neuronal cells makes them attractive tools for tissue repair. While a number of online sites are devoted to "tooth banking" as an alternative to cord blood banking, it is not clear if this technology has reached the level of clinical utility.

VIII. STEM CELL DISEASES

It is possible that stem cell abnormalities are the basis for several diseases and conditions.

Metaplasia is a switch that occurs in tissue differentiation that converts one cell type to another, altering functional ability of a particular cell type in an organ. It is often seen in lung disease (e.g., pulmonary fibrosis) and intestinal disorders (e.g., inflammatory bowel disease and Crohn disease). Metaplasia represents a change that might be brought about by stem cells rather than by terminally differentiated cells.

It is likely that many cancers, particularly of continuously renewing tissues, such as blood, gut, and skin, are in fact diseases of stem cells. Only these cells persist for a sufficient length of time to accumulate the requisite number of genetic changes for malignant transformation (see Chapter 22).

Chapter Summary

- Stem cells can be classified according to their potential for differentiation to different cell types.
- Stem cells are found in embryos and in adult tissues.
- Plasticity refers to the ability to differentiate into different cell types.
- Specific transcription factors play a role in maintaining pluripotency.
- The ability of a cell to maintain stemness and induce differentiation can be controlled by modifying the cell's chromatin.
- Asymmetric division is necessary to maintain stemness.
- The fate of stem cells is governed by intrinsic and extrinsic mechanisms.
- Pluripotency can be induced in adult somatic cells by overexpression of key transcription factors.

Study Questions

Choose the ONE best answer.

1.1 Maintenance and renewal of stem cells require

    A. Chromatin in a relaxed configuration.
    B. Transcription factors in an inhibitory state.
    C. Suppression of signaling mechanisms.
    D. Prevention of entry into the cell cycle.
    E. cell division that generates two daughter stem cells.

Correct answer = A. A relaxed and open configuration of the chromatin allows for regulation by master transcription factors, which mediate inhibition of differentiation and maintenance of stemness. Transcription factors and signaling mechanisms play an important role in regulating this process. Proliferation and an active cell cycle allow renewal and asymmetric cell division ensures stem cell propagation by creating two different daughter cells.

1.2 Master regulators that maintain pluripotency of stem cells are proteins that function as

    A. Actin-binding proteins.
    B. Adhesive proteins.
    C. Microtubule motors.
    D. Second messengers.
    E. Transcription factors.

Correct answer = E. Transcription factors maintain pluripotency by regulating target genes on DNA. Actin-binding proteins are important for assembly and disassembly of actin filaments, a cytoskeletal protein. Cell-cell attachment and attachment of cells to extracellular matrix require cell adhesion molecules. Microtubule motors are proteins that aid in intracellular trafficking. Second messengers relay signals within cells (see Chapters 2, 4, and 17 for details).

1.3 Which of the following is required for maintenance of a stem cell niche?

A. Association of stem cells with only other stem cells
B. Specific extracellular matrix interactions without stem cells
C. Integrin-mediated attachment between stem and other cell types
D. Symmetric separation of fate determining molecules into daughter cells
E. Insensitivity to specific growth factors to stem cells

Correct answer = C. Integrin-mediated attachment between stem cells and other cell types in the stem cell niche. An integrin attaches stem cells to other cell types for maintenance of stemness. In the niche, the stem cells associate with other non–stem cells and are mediated by their interactions with extracellular matrix components. Signaling pathways mediated through growth factors affect the stem cell response. Asymmetric distribution of cellular components is essential for stem cell renewal.

1.4 Induced pluripotency refers to

A. Activation of stemness in embryonic stem cells
B. Generation of cells from the inner cell mass of blastocysts
C. Conversion of a totipotent cell to pluripotency
D. Reprogramming of unipotent cells to be pluripotent
E. The creation of germ cells from the mesoderm

Correct answer = D. Pluripotency can be induced in unipotent cells by the introduction of the core transcription factors that control stem cells. Embryonic stem cells are generated from the inner cell mass and are pluripotent. Totipotent cells can give rise to a whole organism, and germ cells cannot be created from mesoderm, which is a differentiated cell type arising from the blastocyst.

1.5 Transcription factors that maintain pluripotency of stem cells

A. Are activators of specific differentiation pathways.
B. Help to maintain a closed chromatin structure.
C. Inhibit other master transcription factors needed for stemness.
D. Can be overexpressed in somatic cells to create stem cells.
E. Are proteins that are generally absent in totipotent cells.

Correct answer = D. A cocktail of four core transcription factors that are active in stem cells can be used to convert somatic cells to pluripotency. These transcription factors inhibit differentiation and actively regulate an open chromatin. The core transcription factors increase their expression as well as each other in pluripotent cells. Totipotent cells have the capacity to form a whole organism, so they also have active core transcription factors governing totipotency.

Extracellular Matrix and Cell Adhesion

2

I. OVERVIEW

Groups of cells with the same developmental origin are organized into **tissues** and collaborate to perform specific biological functions. There are four basic types of tissue: **epithelial**, **muscle**, **nervous**, and **connective** tissues. Epithelial tissue is found on surface and forms sheets that function in protection, secretion, absorption, and filtration. Muscle tissues are responsible for motion and contraction. Nervous tissue conducts impulses to control muscles, mental activity, and bodily functions. And, the most abundant type of tissue, connective tissue, is widely distributed throughout the body and connects yet separates other tissues from each other. In so doing, connective tissue provides strength, protection, and elasticity. Blood, bone, cartilage, fat, ligaments, lymph, and tendons are examples of connective tissues.

Each tissue creates a stable environment in which its cellular constituents metabolize nutrients, respond to external stimuli, grow, and differentiate according to their own needs and roles within the tissue. Epithelial, muscle, and nerve tissues are composed principally of cells, while connective tissue contains an abundant matrix of macromolecules in its extracellular space. These macromolecules are synthesized and secreted by the cells that reside within the connective tissue and contribute to physical characteristics of the tissue. Together, the macromolecules secreted from the cells of the tissue comprise the **extracellular matrix** (ECM) that contributes to physical characteristics of tissues.

Adhesion is also important for maintaining tissue integrity and for cell-to-cell and cell-to-ECM connections. Structural ECM components within tissues include proteins that mediate adhesion. Transmembrane adhesive proteins within cells also facilitate physical connections between cells and the ECM and serve important roles in cell growth and differentiation and in cellular migration. Cells and the ECM within a tissue influence each other by altering their adhesive connections, creating a complex feedback mechanism.

Tissues, consisting of cells and the ECM and dependent on adhesion, combine to form the recognizable structures of organs, which have very specific functions. For example, muscle and fibrous connective tissue combine to form the heart, whose function is to pump blood. The physical nature and properties of noncellular components of the tissues also contribute to the specialized structures and function of organs. Because a substantial part of tissue volume is extracellular space filled by an intricate network of macromolecules of the ECM, the characteristics and properties of this matrix contribute to organ function. Their proper synthesis is required for normal organ structure and function and for health of the individual person. Altered synthesis of ECM components or damage to proteins and polysaccharides of the ECM can result in disease.

II. EXTRACELLULAR MATRIX

Proteins and polysaccharides synthesized and secreted from cells within a tissue are assembled into an organized, complex network in the ECM. This matrix is specialized to perform different functions in different tissues. For example, the ECM adds strength to tendons and is involved in filtration in the kidney and in attachment in skin.

Cellular and extracellular materials make up different proportions of different tissues (Figure 2.1). While **epithelial tissue** is principally cellular, with neighboring cells attached to one another forming a sheet and containing only a small amount of ECM, **connective tissue** is principally ECM with fewer cells per volume. The ECM of epithelial tissue is known as the **basement membrane** or **basal lamina** and is secreted in the same direction from all epithelial cells in a layer. Therefore, the basement membrane appears beneath the epithelial cells that have secreted it and separates epithelial cells from connective tissue underlying it. The **lamina propria** is a thin layer of loose connective tissue found beneath the epithelial cells or epithelium. Together, the epithelium and the lamina propria constitute the **mucosa.** The terms "mucous membrane" and "mucosa" refer to the epithelium with the lamina propria. **Submucosa** is a layer of tissue beneath a mucous membrane.

The physical nature of the ECM also varies from tissue to tissue. Blood is fluid, while cartilage has a spongy characteristic owing to the nature of extracellular materials in those tissues. Three major categories of extracellular macromolecules make up the ECM: (1) glycosaminoglycans (GAGs) and proteoglycans, (2) fibrous proteins, including collagen and elastin, and (3) adhesive proteins, including fibronectin and laminin. While two of these categories are types of proteins, proteoglycans are mostly carbohydrate in composition.

A. Proteoglycans

Proteoglycans are aggregates of GAGs and proteins. GAGs are also known as **mucopolysaccharides** and are composed of repeating disaccharide chains where one of the sugars is an N-acetylated amino sugar, either N-acetylglucosamine or N-acetylgalactosamine (Figure 2.2), and the other is an acidic sugar. In connective tissue, **ground substance** refers to extracellular gel-like materials deposited between cells and is composed mainly of water and proteoglycans, but not fibrous proteins.

GAGs are organized in long, unbranched chains and adopt extended structures in solution. Most GAGs are sulfated, and all contain multiple negative charges. The most prevalent GAG is chondroitin sulfate. Other GAGs include hyaluronic acid, keratin sulfate, dermatan sulfate, heparin, and heparan sulfate.

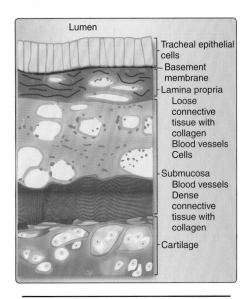

Figure 2.1
Connective tissue underlying an epithelial cell sheet.

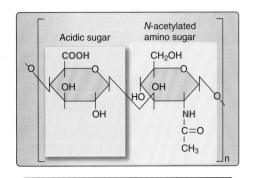

Figure 2.2
Repeating disaccharide unit of GAGs.

1. **Characteristics of proteoglycans:** Because of their net negative surface charges, GAGs repel each other. In solution, GAGs tend to slide past each other, producing the slippery consistency we associate with mucous secretions. Also, because of their negative charges, GAGs attract positively charged sodium ions, which are found in solution complexed with water molecules (Figure 2.3). The hydrated sodium is drawn into the matrix containing GAGs. This flooding of the matrix with water creates swelling pressure (turgor) that is balanced by tension from collagen, a fibrous protein of the ECM. In this manner, GAGs help the ECM resist opposing forces of tissue compression. Cartilage matrix lining the knee joint has large quantities of GAGs and is also rich in collagen. This cartilage is tough and resistant to compression. The bones of the joint are cushioned by the water balloon–like structure of the hydrated GAGs in the cartilage. When compressive forces are exerted on it, the water is forced out and the GAGs occupy a smaller volume (Figure 2.4). When the force of compression is released, water floods back in, rehydrating the GAGs, much like a dried sponge rapidly soaks up water. This change in hydration and ability of the ECM to recover and rehydrate quickly after water has been forced out is referred to as **resilience.** Resilience in the ECM is also observed in synovial fluid and in the vitreous humor of the eye (see also *LIR Biochemistry*, Chapter 14).

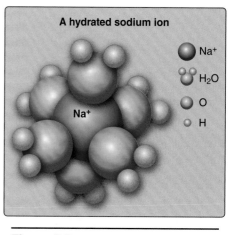

Figure 2.3
Hydration of sodium.

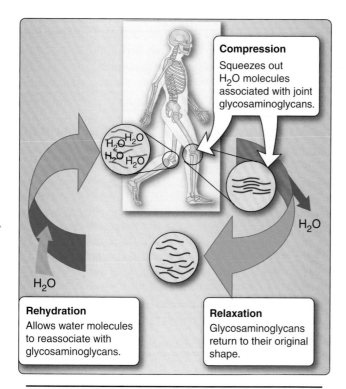

Figure 2.4
Resilience of GAGs.

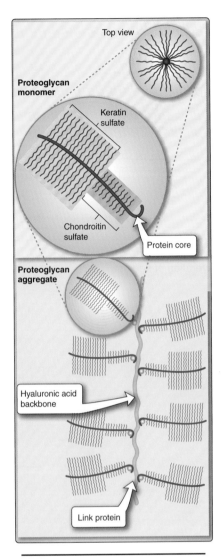

Figure 2.5
Model of cartilage proteoglycan.

2. **Structure of proteoglycans:** With the exception of a very large GAG named hyaluronic acid, other GAGs covalently attach to protein and form **proteoglycan monomers** consisting of a core protein with chains of GAGs extending out from it. In cartilage proteoglycans, the GAGs include chondroitin sulfate and keratin sulfate. Individual GAGs within proteoglycan monomers remain separated from each other owing to charge repulsion. The resulting proteoglycan is often described as having a "bottlebrush" or "fir tree" appearance (Figure 2.5). The individual GAG chains resemble the wire bristles of a brush or the needles on an evergreen tree and the core protein, a branch. The wire central portion of the brush handle or the trunk of the tree is the **hyaluronic acid**. Individual proteoglycan monomers then bind to this large hyaluronic acid GAG, to form a proteoglycan aggregate. The association occurs primarily through ionic interactions between the core protein and the hyaluronic acid and is stabilized by smaller, **link proteins**.

B. Fibrous proteins

The second category of molecules within the ECM are the fibrous proteins. In contrast with globular proteins that have compact structures resulting from secondary, tertiary, or even quaternary protein structure (see *LIR Biochemistry*, Chapter 4), fibrous proteins are extended molecules that have structural functions in tissues. Fibrous proteins are composed of specific types of amino acids in their primary sequence that combine into secondary structural elements but do not have more complex protein structure. **Collagen** and **elastin** are fibrous proteins in the ECM that are important components of connective tissue, skin, and blood vessel walls.

Clinical Application 2.1: Use of Glucosamine as Therapy for Osteoarthritis

Osteoarthritis is the most common form of chronic joint disease worldwide, affecting millions of individuals. Degradation of joint cartilage causes pain, stiffness, and swelling, with progressive worsening of signs and symptoms. Traditional management of the disease has been with drugs that treat the symptoms but do not improve joint health. Glucosamine and chondroitin have been reported both to relieve pain and to stop progression of osteoarthritis. These compounds are readily available as over-the-counter dietary supplements in the United States. Based on several well-controlled clinical studies, it does appear that glucosamine sulfate (but not glucosamine hydrochloride) and chondroitin sulfate may have a small-to-moderate effect in relieving symptoms of osteoarthritis. The European Society for Clinical and Economic Aspects of Osteoporosis and Osteoarthritis (ESCEO) reports that the prescription patented crystalline glucosamine sulfate formulation is superior to other glucosamine sulfate and glucosamine hydrochloride formulations in controlling pain and in having a lasting impact on disease progression. Long-term clinical trials reveal that this form of glucosamine may delay joint structural changes, if treatment is begun early in the management of knee osteoarthritis. Use of prescription patented crystalline glucosamine sulfate for at least 1 year is reported to cause a reduction in the need for total joint replacement for at least 5 years after treatment ends.

1. **Collagen:** Collagen is the most abundant protein in the human body and forms tough protein fibers that are resistant to shearing forces. Collagen is the main type of protein in bone, tendon, and skin. Bundled collagen in tendons imparts strength. In bone, collagen fibers are oriented at an angle to other collagen fibers to provide resistance to mechanical shear stress applied from any direction. Collagen is dispersed in the ECM and provides support and strength.

   Collagen is actually a family of proteins, with 28 distinct types. However, over 90% of collagen in the human body is in collagen types I, II, III, and IV. Together, the collagens constitute 25% of total body protein mass. Types I, II, and III are fibrillar collagens whose linear polymers of fibrils reflect the packing together of individual collagen molecules. Type IV (and also type VII) is a network-forming collagen that becomes a three-dimensional mesh rather than distinct fibrils.

   a. **Structure of collagen:** Collagen molecules are composed of three helical polypeptide α **chains** of amino acids that wind around one another forming a collagen triple helix (Figure 2.6). The various types of collagen have different α chains, occurring

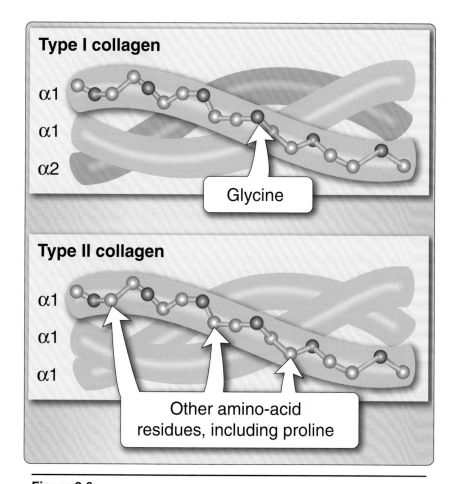

Figure 2.6
Triple helical structure of collagen.

Table 2.1: Chain Compositions of Collagen Types

Type	Chain Composition	Characteristics
I	$[\alpha_1(I)]_2[\alpha_2(I)]$	**Most abundant collagen** Found in bones, skin, tendons Present in scar tissue
II	$[\alpha_1(II)]_3$	Found in hyaline cartilage Present on the ventral ends of the ribs, in the larynx, trachea, and bronchi, and on the articular surface of bone
III	$[\alpha_1(III)]_3$	Collagen of granulation tissue in healing wounds Produced before tougher type I is made Forms reticular fibers Found in artery walls, intestine, uterus
IV	$[\alpha_1(IV)]_2[\alpha_2(IV)]$	Found in basal lamina and eye lens Part of the filtration system in glomeruli of nephron in kidneys

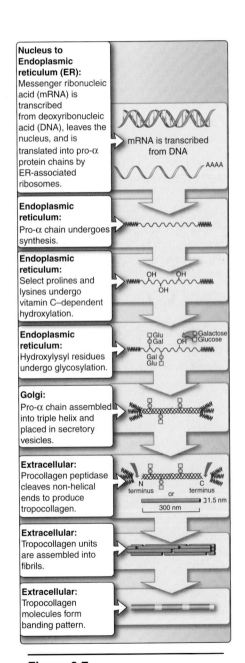

Nucleus to Endoplasmic reticulum (ER):
Messenger ribonucleic acid (mRNA) is transcribed from deoxyribonucleic acid (DNA), leaves the nucleus, and is translated into pro-α protein chains by ER-associated ribosomes.

Endoplasmic reticulum:
Pro-α chain undergoes synthesis.

Endoplasmic reticulum:
Select prolines and lysines undergo vitamin C–dependent hydroxylation.

Endoplasmic reticulum:
Hydroxylysyl residues undergo glycosylation.

Golgi:
Pro-α chain assembled into triple helix and placed in secretory vesicles.

Extracellular:
Procollagen peptidase cleaves non-helical ends to produce tropocollagen.

Extracellular:
Tropocollagen units are assembled into fibrils.

Extracellular:
Tropocollagen molecules form banding pattern.

Figure 2.7
Synthesis of collagen.

in distinct combinations (Table 2.1). For example, collagen type I has two type I α_1 chains and one type I α_2 chain, while collagen type II has three type II α_1 chains. In the primary amino acid sequence of α chains, every third amino acid is **glycine** (Gly), whose side chain is simply a hydrogen atom. Collagen is also rich in the amino acids **proline** (Pro) and **lysine** (Lys). The amino acid sequence of most of the α chains can be represented as repeating units of -X-Y-Gly—where X is typically Pro and Y is often a modified form of either Pro or Lys (hydroxyproline [Hyp] or hydroxylysine [Hyl]). In the collagen triple helix, the small hydrogen side chains of Gly residues (residues are amino acids within proteins) are placed toward the interior of the helix in a space too small for any other amino acid side chain. Three α chains in this conformation can pack together tightly. Pro also facilitates the formation of the helical conformation of each α chain because it has a ring structure that causes "kinks" in the peptide chain. (See also *LIR Biochemistry*, Chapter 1, pp. 2–3, for structures of amino acids.)

b. **Synthesis of collagen:** Individual fibrillar collagen polypeptide chains are translated on membrane-bound ribosomes (Figure 2.7). Then, an unusual modification is made to certain Pro and Lys amino acid residues. In reactions that require molecular oxygen, Fe^{2+} and the reducing agent **vitamin C** (ascorbic acid), prolyl hydroxylase and lysyl hydroxylase catalyze the **hydroxylation** (addition of OH) of selected Pro and Lys (prolyl and lysyl residues) within the new protein (Figure 2.8). Some of the hydroxylysyl residues are then **glycosylated** (carbohydrate is added to them).

In the Golgi complex, three pro-α chains assemble into a helix by a zipper-like folding. A secretory vesicle then buds off from the Golgi complex, joins with the plasma membrane, and releases the newly synthesized collagen triple helices into the extracellular space. Propeptides, small portions of each end (C-terminal and N-terminal) of the newly synthesized chains, are cleaved by proteases (such as procollagen peptidase)

Clinical Application 2.2: Collagen and Aging

Because collagen plays a key role in the support structure of the skin, if collagen production slows or its structure is changed, the appearance of the skin will also change. As skin ages, collagen production slows down. Additionally, over time, collagen fibers become rigid. Theoretically, this damage is caused by free radicals that attach to collagen, causing collagen strands to bind together and become thicker and more unyielding to movement. This process is the basis for wrinkle formation and the loss of firmness in mature skin. Collagen injections are sometimes given to restore volume and reduce the appearance of wrinkles. Antiaging products commonly contain antioxidants that may inhibit free radicals and slow the damage to collagen. Some other OTC topical products contain plant or animal collagen, in the hope that it will be absorbed by the skin and replenish the natural collagen that has been lost due to aging. More promising topical products may contain retinoids that inhibit synthesis of collagenases, the enzymes that break down collagen. Retinoic acid, available in prescription creams, can also stimulate the production of new collagen fibers in the skin. While hundreds of OTC antiaging products are available, they contain either much lower concentrations or different active ingredients than creams available by prescription. The efficacy of OTC antiaging products is not proven.

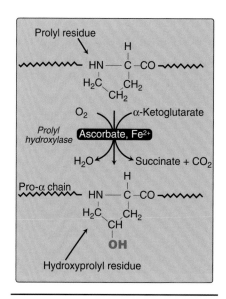

Figure 2.8

Hydroxylation of proline residues of pro-α chains of collagen by prolyl hydroxylase. (From Ferrier D. R. (2014). *Biochemistry* (6th ed.). Philadelphia, PA: Wolters Kluwer.)

to yield tropocollagen. Self-assembly and cross-linking of **tropocollagen** molecules then occur to form mature **collagen fibrils**. Packaging of collagen molecules within fibrils leads to a characteristic repeating structure with a banding pattern that can be observed with electron microscopy.

The fibrillar array of collagen is acted on by the enzyme, **lysyl oxidase**, which modifies lysyl and hydroxylysyl amino acid residues, forming **allysine** (allysyl residues) and allowing them to form the covalent cross-links seen in mature collagen fibers (Figure 2.9). This cross-linking is essential to achieve the tensile strength necessary for proper functioning of connective tissue.

2. **Elastin:** The other major fibrous protein in the ECM is elastin. Elastic fibers formed by elastin enable skin, arteries, and lungs to stretch and recoil without tearing. Elastin is rich in the amino acids glycine, alanine, proline, and lysine. Similar to collagen, elastin contains hydroxyproline, although only a small amount. No carbohydrate is found within the structure of elastin, and therefore, it is not a glycoprotein.

    a. **Synthesis of elastin:** Cells secrete the elastin precursor, tropoelastin, into the extracellular space. Tropoelastin then interacts with glycoprotein microfibrils including **fibrillin**, which serve as a scaffolding onto which tropoelastin is deposited. Side chains of some of the lysyl residues within tropoelastin polypeptides are modified to form **allysine**. In the next step, the side chains of three allysyl residues and the side chain of one unaltered lysyl residue from the same or neighboring tropoelastin polypeptide are joined covalently to form a **desmosine cross-link**

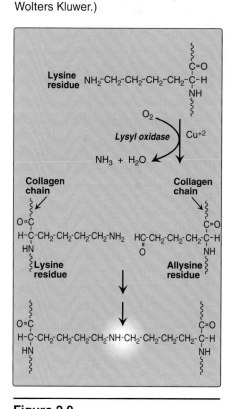

Figure 2.9

Formation of cross-links in collagen. (Note: *Lysyl oxidase* is irreversibly inhibited by a toxin from plants in the genus Lathyrus, leading to a condition known as lathyrism.) (From Ferrier, D. R. (2014). *Biochemistry* (6th ed.). Philadelphia, PA: Wolters Kluwer.)

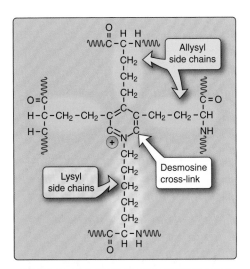

Figure 2.10
Desmosine cross-link in elastin.

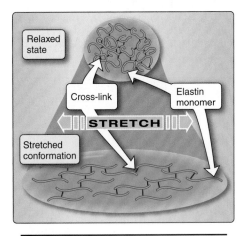

Figure 2.11
Elastin in relaxed and stretched conformations.

(Figure 2.10). Thus, four individual polypeptide chains are covalently linked together.

   b. **Characteristics of elastin:** The structure of elastin is that of an interconnected rubbery network that can impart stretchiness to the tissue that contains it. This structure resembles a collection of rubber bands that have been knotted together, with the knots being the desmosine cross-links. Elastin monomers appear to lack an orderly secondary protein structure because elastin can adopt different conformations both when relaxed and when stretched (Figure 2.11).

C. Fibrous proteins and disease

Because collagen and elastin play important structural roles in tissues, impaired production of these fibrous proteins can result in disease states. Both acquired and inherited defects can result in abnormal fibrous proteins that change the physical properties of the tissue, sometimes with serious consequences. In other situations, the fibrous proteins are synthesized properly but are then degraded inappropriately, also impacting the normal functional characteristics of the tissue. The disease states described below serve to highlight the importance of normal fibrous proteins to the health of tissues and of the individual.

1. **Scurvy:** Dietary **deficiency in vitamin C** causes scurvy, a disorder that results from aberrant collagen production. In the absence of vitamin C, hydroxylation of prolyl and lysyl residues cannot occur, resulting in defective pro-α chains that cannot form a stable triple helix. These abnormal collagen pro-α chains are degraded within the cell. As a consequence, there is less normal, functional collagen available to replace collagen that has reached the end of its functional life span. Therefore, there is less collagen present to provide strength and stability to tissues. Blood vessels become fragile, bruising occurs, wound healing is slowed, and gingival hemorrhage and tooth loss occur (Figure 2.12A).

2. **Osteogenesis imperfecta:** In contrast to the acquired collagen deficiency of scurvy, inherited collagen defects impact individuals throughout their life span, as opposed to only after a dietary

---

Clinical Application 2.3: Scurvy: Past and Present

In centuries past, scurvy was a ravaging illness among sailors who spent many months out at sea. They began to carry limes onboard ships to provide a source of vitamin C during long voyages. While much less common in the 21st century, scurvy does still occur today, even in industrialized societies. It is seen predominantly among indigent persons, elderly individuals who live alone and prepare their own food, persons with poor dentition, persons with alcoholism, and those who follow fad diets. Other individuals may avoid fruits and vegetables, dietary sources of vitamin C, because of perceived food allergies or intolerances. Scurvy is less common in the pediatric population. But, signs of scurvy in children may mimic child abuse, since developing bones may demonstrate effects of scurvy more than adult bones. Scurvy may actually occur as a result of parental neglect in not providing proper foods to the children. Because vitamin C deficiency today is typically accompanied by deficiencies in other essential nutrients, diagnosis may be delayed.

deficiency began. A family of inherited collagen disorders, osteogenesis imperfecta has been called "brittle bone disease," because many affected individuals have weak bones that fracture easily (Figure 2.12B). The disorder is caused by any one of several inherited mutations in a collagen gene, resulting in weak bones. A mutation may result in decreased production of collagen or in abnormal collagen. Eight forms of osteogenesis imperfecta are currently known. Some forms have more severe signs and symptoms than others (Table 2.2). The most common is osteogenesis

Table 2.2: Osteogenesis Imperfecta Types and Characteristics

Type	Inheritance	Collagen	Features
I	Autosomal dominant	Normal structure Low concentration	Bones fracture easily, most before puberty Normal stature Loose joints, weak muscles Blue/gray-tinted sclera Bone deformity absent Triangular face Brittle teeth possible Hearing loss possible in 20s and 30s
II	Autosomal dominant	Abnormal structure Collagen I mutation	Lethal at or shortly after birth Numerous fractures Small stature Severe bone deformity Respiratory problems Underdeveloped lungs
III	Autosomal dominant	Abnormal structure Collagen I mutation	Bones fracture easily Fractures often present at birth Short stature Blue/gray-tinted sclera Loose joints and poor muscle development Barrel-shaped rib cage Triangular face Spinal curvature Respiratory problems Brittle teeth Hearing loss possible
IV	Autosomal dominant	Abnormal structure Collagen I mutation	Bones fracture easily, most before puberty Shorter-than-average stature Sclera normal in color Mild-to-moderate bone deformity Tendency toward spinal curvature Triangular face Brittle teeth possible Hearing loss possible
V	Autosomal dominant	Abnormal structure Normal collagen I Unknown mutation	Clinically similar to type IV Histologically bone is "mesh-like"
VI	Unknown autosomal Eight reported cases	Abnormal structure Normal collagen I Unknown mutation	Clinically similar to type IV Histologically, bone ham "fish-scale" appearance
VII	Recessive	Abn ormal collagen CRTAP gene mutation	Resembles type IV or lethal type II
VIII	Recessive	Abnormal collagen LEPRE1 gene mutation	Resembles lethal type II or type III Severe growth deficiency Extreme skeletal undermineralization

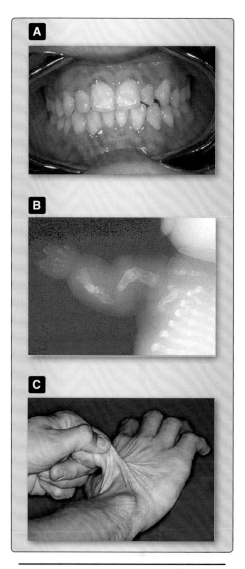

Figure 2.12
Signs of conditions with abnormal fibrous proteins. **A.** Bleeding gums of a patient with scurvy. **B.** Fractured bone of a patient with osteogenesis imperfecta. **C.** Stretchy skin of an individual with Ehlers-Danlos syndrome.

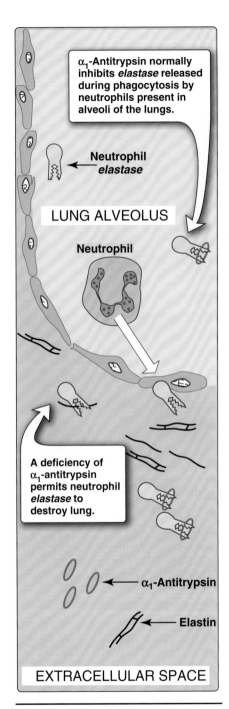

α₁-Antitrypsin normally inhibits *elastase* released during phagocytosis by neutrophils present in alveoli of the lungs.

Neutrophil *elastase*

LUNG ALVEOLUS

Neutrophil

A deficiency of α₁-antitrypsin permits neutrophil *elastase* to destroy lung.

α₁-Antitrypsin

Elastin

EXTRACELLULAR SPACE

Figure 2.13
Destruction of alveolar tissue by *elastase* released from neutrophils activated as part of the immune response to airborne pathogens. (From Ferrier, D. R. (2014). *Biochemistry* (6th ed.). Philadelphia, PA: Wolters Kluwer.)

imperfecta type I, in which most affected persons have mild signs and symptoms including bones that fracture easily, especially prior to puberty. They may also have spinal curvature and hearing loss. By contrast, type II is lethal prior to or shortly after birth. Most types have autosomal dominant inheritance patterns and affected persons inherit a mutant gene from one parent who is also affected.

3. **Ehlers-Danlos syndrome:** Ehlers-Danlos syndrome (EDS) is a group of disorders that generally result from inherited defects in the structure, production, or processing of fibrillar collagen. There are six primary subtypes as well as some other recognized more rare types of EDS. Hypermobility followed by the classical forms of EDS are most common. The types are categorized based on signs and symptoms. Most are inherited as autosomal dominant traits. Abnormally flexible and loose joints that extend beyond the normal range of motion and stretchy but excessively fragile skin and blood vessels are characteristics (Figure 2.12C).

4. **Marfan syndrome:** Another condition inherited as an autosomal dominant trait is Marfan syndrome. In this disorder, a mutation occurs in the gene that codes for the **fibrillin**-1 protein essential for maintenance of elastin fibers. Because elastin is found throughout the body and is particularly abundant in the aorta, the ligaments, and in portions of the eye, these sites are most affected in individuals with Marfan syndrome. Many affected individuals have ocular abnormalities and myopia (near-sightedness) and abnormalities in their aorta. They also have long limbs and long digits, tall stature, scoliosis (side-to-side or front-to-back spinal curvature) or kyphosis (curvature of the upper spine), abnormal joint mobility, and hyperextensibility of hands, feet, elbows, and knees.

5. **α₁-Antitrypsin deficiency:** Also related to elastin is α₁-antitrypsin deficiency (Figure 2.13). This is an autosomal dominant disorder caused by deficiency of the protease inhibitor that normally regulates actions of elastase, an enzyme that degrades elastin. In the lungs of all individuals, the alveoli (small air sacs) are chronically exposed to low levels of neutrophil elastase, released from activated neutrophils. This destructive enzyme is, however, normally inhibited by α₁-antitrypsin, also called α₁-protease inhibitor and α₁-antiprotease, the most important physiological inhibitor of neutrophil elastase. Individuals deficient in α₁-antitrypsin have reduced ability to inhibit elastase in the lung. Because the lung tissue cannot regenerate, the destruction of the connective tissues of alveolar walls is not repaired and disease results. Affected individuals usually present with symptoms of lung disease between the ages of 20 and 50. At first, they may experience shortness of breath after mild physical activity, but the condition often progresses to emphysema, where the alveoli (small air sacs) of the lungs are irreversibly damaged. Tobacco smoke accelerates lung damage in affected individuals. In the United States, it is estimated that 2% to 5% of patients with emphysema have an inherited defect in α₁-antitrypsin (see also *LIR Biochemistry*, Chapter 4).

D. Adhesive proteins

The last category of ECM components consists of proteins that join together and organize the ECM and also link cells to the ECM.

Fibronectin and **laminin** are adhesive glycoproteins secreted by cells into the extracellular space. Fibronectin is the principal adhesive protein in connective tissues, while laminin is the principal adhesive protein in epithelial tissues. Both are considered multifunctional proteins because they contain three different binding domains that link them to cell surfaces and to other components of the ECM, including proteoglycans and collagen (Figure 2.14). Through their interactions with fibronectin or laminin, proteoglycans and collagen are linked to each other and to a cell's surface. Thus, adhesive proteins join ECM components to each other and link cells to the ECM.

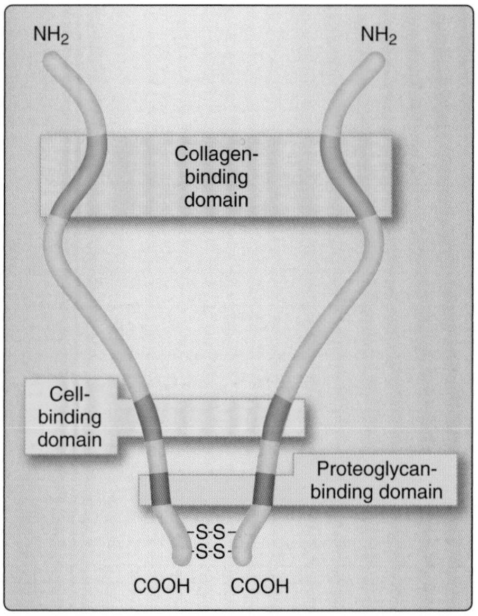

E. ECM degradation and remodeling

The ECM is highly dynamic and undergoes remodeling where ECM components are deposited, degraded, and modified. Remodeling allows for processes that include regulation of cell differentiation, establishment of stem cell niches, wound repair, and bone remodeling. Enzymes involved in ECM remodeling include families of metalloproteinases, the matrix metalloproteinase (**MMP**), transmembrane proteases known as **ADAMs** (a disintegrin and metalloproteinases), and the related, secreted proteases, **ADAMTS**s (ADAM with thrombospondin domain). Some serine proteases also degrade ECM adhesive protein ECM components. While some MMPs target ECM components including proteoglycans and adhesive proteins, others degrade collagen. Several members of the ADAMTS family degrade proteoglycans, while others are involved in collagen remodeling. Tissue inhibitors of metalloproteinases (**TIMPs**) normally regulate the function of specific metalloproteinases.

If the ECM is not remodeled properly and tissue dynamics are altered, it can lead to pathological consequences. Overexpression of MMPs as well as mutations that cause lack of function of MMPs or ADAMTS can be causes of altered ECM remodeling. Consequences can include alterations in cell differentiation, cell proliferation, and cell death, as well as pathological processes that may include tissue fibrosis and cancer. The pathological destruction of cartilage and bone in joints in patients with osteoarthritis and rheumatoid arthritis occurs in part from overexpression of metalloproteinases.

III. CELL ADHESION

Cell-to-ECM and cell-to-cell adhesions are mediated by plasma membrane–anchored proteins called **cell adhesion molecules.** Collections of adhesion molecules form **cell junctions** that join cells together within tissues. There is an increasing recognition that adhesion is involved in the pathogenesis of many different diseases including viral infections, cardiovascular disease, and bone and joint disease. Development of a more complete understanding of the fundamental process of cell adhesion will likely lead us to a better understanding of such diverse pathologies.

A. Adhesion in developing tissues

Many tissues, including most epithelial tissues, develop from a precursor, the founder cell that divides to produce copies of itself. These newly produced cells remain attached to the ECM and/or to other cells owing to cell adhesion (Figure 2.15). A growing tissue is able

Figure 2.14
Structure of a fibronectin dimer.

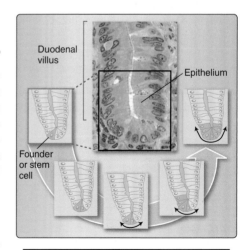

Figure 2.15
Cell adhesion during epithelial tissue development.

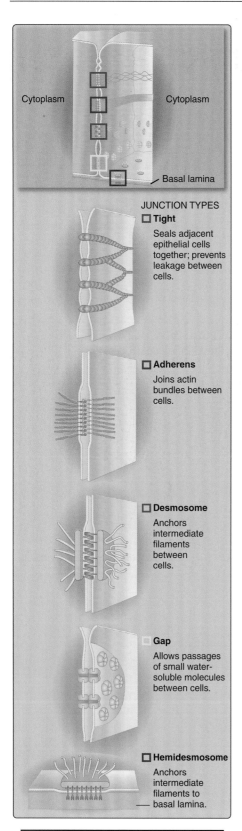

JUNCTION TYPES

☐ **Tight**

Seals adjacent epithelial cells together; prevents leakage between cells.

☐ **Adherens**

Joins actin bundles between cells.

☐ **Desmosome**

Anchors intermediate filaments between cells.

☐ **Gap**

Allows passages of small water-soluble molecules between cells.

☐ **Hemidesmosome**

Anchors intermediate filaments to basal lamina.

Figure 2.16
Types of cell junctions.

to form because the member cells remain attached and do not travel elsewhere. Selective adhesion is essential for the development of tissues that have complex origins. Migration of cells is also needed in these situations. One population of cells invades another and selectively adheres to them and, possibly, to other cell types to assemble the tissue.

B. Cell junctions

Even in mature tissues, structure and stability are actively maintained by selective cell adhesions. These adhesions are formed by the cells and fine-tuning and adjustment of them are ongoing. Cells in tissues adhere to other cells at specialized regions known as **cell junctions**, which are classified according to their function (Figure 2.16). For example, physical barriers between individual cells are formed with **tight or occluding junctions** (also called zonulae occludentes). **Desmosomes or anchoring junctions** (also called maculae adherentes) as well as **adherens junctions** (also called zonulae adherentes) function to couple neighboring cells to each other by interacting with components of the cytoskeleton (intermediate filaments and actin), the inner framework, or scaffolding within cells (see Chapter 4). **Hemidesmosomes** link cytoskeletal intermediate filaments to the basal lamina. And finally, **gap or communicating junctions** (also called nexus) allow for the transfer of signals between cells. Cell junctions are important in maintaining the structure of a tissue as well as its integrity. They are composed of a collection of individual cell adhesion molecules.

C. Cell adhesion molecules

Cell adhesion molecules mediate selective cell-to-cell and cell-to-ECM adhesion. These are all **transmembrane proteins** that are embedded within the plasma membranes of cells. They extend from the cytoplasm through the plasma membrane to the extracellular space. In the extracellular space, they bind specifically to their ligands. The ligands may be cell adhesion molecules on other cells, certain molecules on the surface of other cells, or components of the ECM. Interactions between individual adhesion molecules are important in adhesion during development and also mediate for cell migration. Four families of adhesion molecules function in cell-cell adhesion: the **cadherins**, the **selectins**, **the immunoglobulin superfamily**, and the **integrins** (Table 2.3). The integrins also function in cell-to-ECM adhesion (see *LIR Immunology*, Chapter 13).

1. **Cadherins:** The cell adhesion molecules that are important in holding cells together to maintain the integrity of a tissue are the cadherins (Figure 2.17A). These transmembrane linker proteins contain extracellular domains that bind to a cadherin on another cell. Cadherins also have intracellular domains that bind to linker proteins of the **catenin** family that bind to the actin cytoskeleton, the internal scaffolding in the cytoplasm (see Chapter 4). Therefore, when two cells are linked together via cadherins, their internal actin cytoskeletons are indirectly linked as well. **Calcium** is required for cadherin binding to another cadherin. Adhesion mediated by cadherins is long-lasting and important in maintaining the tissue structure.

Table 2.3: Adhesion Molecules and Ligands

Family	Name	Synonym(s)	Expressed By	Ligand(s)
Cadherins	**Classical**			
	E-cadherin N-cadherin P-cadherin	CDH1 CDH2 CDH3	Epithelial tissue Neurons Placenta	E-cadherin N-cadherin P-cadherin
	Desmosomal			
	Desmocollins Desmogleins	DSC1, 2, 3 DSG1, 2, 3	Epithelial tissue Epithelial tissue	Desmocollins Desmogleins
Selectins	E-selectin	CD62E	Activated endothelium	Sialyl Lewis X
	L-selectin	CD62L	Leukocytes	CB34 GlyCAM-1 MadCAM-1 Sulfated sialyl Lewis X
	P-selectin	CD62P	Platelets, activated endothelium	Sialyl Lewis X, PSGL-1
Immunoglobulin superfamily	CD2	LFA-2	T cells	LFA-3
	ICAM-1	CD54	Activated endothelium, lymphocytes, dendritic cells	LFA-1 Mac-1
	ICAM-2	CD102	Dendritic cells	LFA-1
	ICAM-3	CD50	Lymphocytes	LFA-1
	LFA-3	CD58	Antigen-presenting cells, lymphocytes	CD2
	VCAM-1	CD106	Activated endothelium	VLA-4
Integrins	LFA-1	CD11a:CD18	Phagocytes, neutrophils, T cells	ICAM-1, -2, -3
	Mac-1	CD11b:CD18	Neutrophils, macrophages, monocytes	ICAM-1 iC3b Fibrinogen
	CR4	CD11c:CD18	Dendritic cells, neutrophils, macrophages	iC3b
	VLA-4	CD49d:CD29	Lymphocytes, macrophages, monocytes	VCAM-1

2. **Selectins:** Some other adhesion molecules mediate more transient cell-to-cell adhesions. For example, selectins are particularly important in the immune system in mediating white blood cell migration to sites of inflammation. Selectins are named for their "**lectin**" or carbohydrate-binding domain in the extracellular portion of their structure (Figure 2.17B). A selectin on one cell interacts with a carbohydrate-containing ligand on another cell.

3. **Immunoglobulin superfamily:** Another family of cell-to-cell adhesion molecules is named because they share structural characteristics of the immunoglobulins (antibodies). Adhesion molecules that are members of the immunoglobulin superfamily fine-tune and regulate cell-to-cell adhesions (Figure 2.17C). Some immunoglobulin superfamily members facilitate adhesion of leukocytes to endothelial cells lining the blood vessels during injury and stress. Ligands for this family of adhesion molecules include other members of the immunoglobulin superfamily as well as integrins.

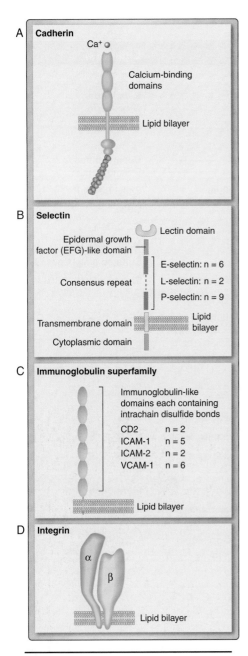

Figure 2.17
Adhesion molecule structure.
A. Cadherin. **B.** Selectin.
C. Immunoglobulin superfamily.
D. Integrin.

4. **Integrins:** Integrins are cell adhesion molecules that can mediate both cell-to-cell and cell-to-ECM adhesions. Members of this family of homologous transmembrane, heterodimeric proteins bind to their ligands with relatively low affinity; multiple weak adhesive interactions characterize integrin binding and function. Integrins consist of two transmembrane chains, α and β (Figure 2.17D). At least 19 α and 8 β chains are known at present. Different α and β chains combine to give integrins with distinct binding properties. The β_2-type subunit is expressed exclusively by leukocytes (white blood cells).

a. **Ligands:** When integrins mediate cell-to-cell adhesions, their ligands are members of the immunoglobulin superfamily. When integrins join a cell to the ECM, collagen and fibronectin commonly serve as their ligands. The cell-binding domain of a fibronectin molecule is its integrin-binding site. The extracellular domains of integrins bind to components of the ECM through recognition of a group of three amino acid residues: arginine, glycine, and aspartic acid, known as an **RGD tripeptide** (based on the one-letter abbreviations for each of the three amino acids). This binding triggers changes in the intracellular cytoplasmic domains of integrins, altering their interaction with cytoskeletal and/or other proteins that regulate cell adhesion, growth, and migration. The intracellular portions of most integrins are joined to bundles of cytoskeletal actin filaments. Thus, integrins mediate interactions between the cytoskeleton within the cell and the ECM surrounding the cell.

b. **Signaling:** Signals generated inside the cell can alter the activation state of some integrins, affecting their affinity for their extracellular ligands. Thus, integrins have a unique ability to send signals across the plasma membrane in both directions, a process referred to as **inside-out and outside-in signaling**.

D. Adhesion and disease

Normal expression and function of adhesion molecules are required to maintain health and to defend against disease. When these normal cell-to-cell and/or cell-to-matrix interactions are interrupted or altered, disease processes can be triggered. Trafficking or movement of immune cells to a site of inflammation within a tissue depends upon adhesion molecules on leukocytes as well as on the endothelium. Impaired adhesion molecule expression causes an interruption in this process. However, adhesion molecules can also be exploited by infectious agents and disease processes. Increased adhesion molecule expression can contribute to inflammatory conditions including asthma and rheumatoid arthritis.

1. **Extravasation (cell migration from circulation to tissue):** When a leukocyte from the immune system responds to an infectious agent in a tissue, its adhesion molecules must encounter their ligands and facilitate that cell's movement from blood into tissue (see also *LIR Immunology*, Figure 13.3).

a. **Steps:** In this process, a selectin on the leukocyte binds to its ligand, often a member of the immunoglobulin superfamily on the surface of an endothelial cell. "**Rolling**" of the leukocyte along the endothelium of the blood vessel then ensues (Figure 2.18). **Activation** of an integrin on the same leukocyte occurs, in an inside-out fashion, owing to signaling set off by the selectin interacting with its ligand. The activated integrin can then bind to its ligand on the endothelium, causing a **firm arrest** of the leukocyte. This is followed by **diapedesis**, or movement through the endothelial layer, and **extravasation**, or entry of the leukocyte into the tissue. Understanding of these traditional three steps of rolling, activation, and firm binding has recently been augmented and refined. Slow rolling, adhesion strengthening, intraluminal crawling, and paracellular and transcellular migration are now recognized as separate, additional steps.

b. **Fatty streak formation:** This same general process that allows cells from the immune system to reach a tissue site of infection also occurs in **fatty streak formation**, one of the first pathological changes in cardiovascular disease. The atherosclerotic process begins with an injury to the inner lining of the blood vessel, the endothelium. Monocytes bind to the injured endothelium in an adhesion molecule–dependent manner and then undergo diapedesis and extravasation into the subendothelium.

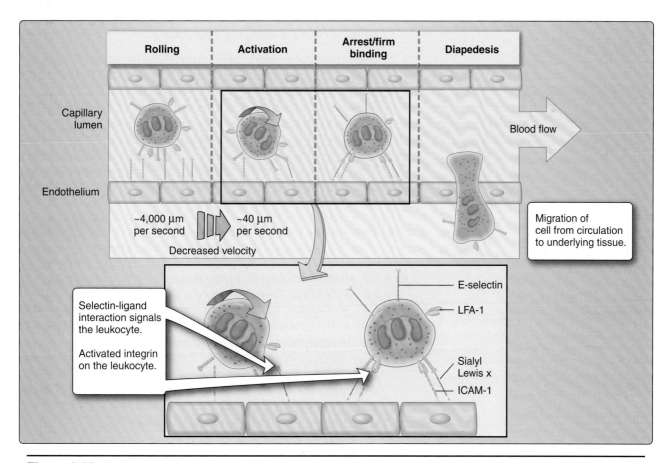

Figure 2.18
Extravasation.

There they engulf excess lipids to become foam cells. The foam cells accumulate in the wall of the blood vessel–forming plaque that becomes calcified. Restriction of blood flow can occur as a consequence (see also *LIR Biochemistry*, p. 235).

2. **Adhesion molecule defects:** Impaired expression of particular adhesion molecules can inhibit leukocyte trafficking to the sites of infection. Abnormal expression of other adhesion molecules can result in the disruption of normal tissue structure. In both situations, the health of the individual is compromised.

   a. **Epithelial-mesenchymal transition (EMT):** EMT occurs when epithelial cells undergo changes in adhesion and polarity. While this is seen in embryogenesis, it is also a common feature of cancer progression. EMT allows the cancers to gain migratory and invasive properties. Changes in the expression or function of adhesion molecules are thought to be involved in the progression of cancer. Most cancers originate from epithelial tissue and E-cadherin is critically important in organizing the epithelium. The function of E-cadherin is altered in most epithelial tumors. Studies have shown that this loss of E-cadherin–mediated cell-to-cell adhesion occurs during tumor progression and is also required for subsequent tumor spreading or metastasis. Since the primary cause of death in persons with cancer is metastatic dissemination of tumor cells, much research is focused on understanding the molecular mechanisms of cell adhesion and signaling in this process.

   b. **Leukocyte adhesion deficiency:** The importance of functional adhesion molecules to health is highlighted by **leukocyte adhesion deficiency I** (LAD), a rare but significant immunodeficiency. In contrast to changes in adhesion molecules later in life that may occur in cancer progression, LAD is an inherited defect in the β_2 **subunit of integrins**, which is normally exclusively expressed on leukocytes. Therefore, their leukocytes have an impaired ability to traffic to the sites of infection and recurrent bacterial infections result. Persons with LAD generally do not survive beyond 2 years of age.

   c. **Pemphigus:** Another disease involving adhesion molecule defects is pemphigus, in which blisters develop as a result of failed cell-to-cell adhesion. Pemphigus is an autoimmune condition characterized by the disruption of cadherin-mediated cell adhesions. There are three types of pemphigus, which vary in severity. All forms are caused by autoantibodies that bind to the proteins in a subfamily of the cadherins, known as the **desmogleins**. Antibody binding to desmogleins prevents their function in cell adhesion. Therefore, adjacent epidermal cells are unable to adhere to each other and blisters develop. (**Pemphigoid** is a related group of blistering conditions in which autoantibodies to proteins of hemidesmosomes impair cell attachment to the underlying basal lamina.)

Clinical Application 2.4: Forms of Pemphigus

Of the three forms of pemphigus, pemphigus vulgaris is the most common and is characterized by mouth sores. Pemphigus foliaceus is the least severe. It is characterized by crusty sores on the scalp, chest, back, and face and is often misdiagnosed as eczema or dermatitis. The least common and most severe form of pemphigus is the malignant variety, known as paraneoplastic pemphigus. It is usually found in conjunction with another malignancy. Very painful sores appear on the mouth, lips, and esophagus. Fatal destruction of alveoli in lung tissue may also occur in this form.

3. **Increased adhesion molecule expression and inflammation:** Expression of more than the usual number of adhesion molecules per cell can result in enhanced migration of cells to a region and can lead to inappropriate inflammation.

   a. **Asthma:** If inappropriate inflammation becomes chronic, as in asthma, discovering the source of the ongoing inflammation is an important step in prevention. In asthma, the airway constricts and becomes inflamed. Attacks are often triggered by viral infections. ICAM-1, a member of the immunoglobulin superfamily that normally facilitates adhesion between endothelial cells and leukocytes after injury or stress, has been implicated in the pathogenesis of asthma. Increased ICAM-1 expression is observed in the respiratory tract of individuals with asthma. This may permit an inappropriately large number of immune cells to migrate there, stimulating chronic inflammation.

   b. **Rheumatoid arthritis:** The pathogenesis of another inflammatory condition, rheumatoid arthritis, may also include increased expression of adhesion molecules. In this autoimmune disease, bone cells may have increased expression of adhesion molecules. In rheumatoid arthritis, synovial inflammation is associated with increased leukocyte adhesion. Selective involvement of the integrin LFA-1 and of ICAM-2 has been demonstrated. Inhibition of certain adhesion molecules is a potential therapy for rheumatoid arthritis.

4. **Adhesion molecules as receptors for infectious agents:** Adhesion molecules may assist in inflammation and in infection in yet another way. Because adhesion molecules are widely expressed on human cells and because viruses need a host binding protein to initiate infection, adhesion molecules may sometimes serve in this role. The same ICAM-1 molecule that mediates attachment of leukocytes to endothelial cells is also used as a receptor by the major group of rhinoviruses, the most important etiologic agent of common colds. Rhinovirus infections are also a major cause of exacerbations in asthma. Blocking ICAM-1 may be a therapeutic method of inhibiting rhinovirus infections.

Chapter Summary

- Tissues are composed of cells and extracellular macromolecules that are produced by the cells of the tissue. A substantial portion of tissue volume is filled by ECM.
- ECM contains proteoglycans, fibrous proteins, and adhesive proteins.
- Proteoglycans are composed of glycosaminoglycans and small, link proteins. They provide resilience and resistance to compressive forces.
- Fibrous proteins include collagen and elastin. Collagen forms tough fibers resistant to shearing stress. Elastin enables tissues to stretch and recoil without tearing.

Abnormalities in fibrous proteins result in disease:

  - Scurvy—impaired collagen production caused by lack of dietary vitamin C.
  - Osteogenesis imperfecta—inherited disorders of collagen characterized by weak bones.
  - Ehlers-Danlos syndrome—characterized by hyperextendible joints and stretchy skin.
  - Marfan syndrome—defective fibrillin-1 impairs the maintenance of elastin and results in defects in the aorta, eye, and skeleton.
  - α_1-Antitrypsin deficiency—predisposes individuals to emphysema. Proteolytic effects of elastase on elastin are unopposed when insufficient α_1-antitrypsin is present.
- The ECM undergoes continual remodeling, regulated by activity of matrix metalloproteinases.
- Cell adhesion is required for normal tissue structure.
- Cell junctions are composed of adhesion molecules that mediate cell-to-cell and cell-to-ECM adhesion. **Families of cell adhesion** molecules include the following:
  - Cadherins—bind to cadherins on other cells to provide long-lasting adhesion between cells in tissues.
  - Selectins—bind to carbohydrate-containing ligands on other cells and mediate movement of leukocytes.
  - Immunoglobulin superfamily members—fine-tune and regulate cell-to-cell adhesions.
  - Integrins—mediate both cell-to-cell and cell-to-ECM adhesions.
- Disruptions in cell adhesion can result in disease.
- Extravasation—the normal process of movement of a leukocyte into a tissue site of infection can also be used by monocytes in fatty streak formation.
- Changes in adhesion molecule expression may be involved in the progression of cancer.
- Deficiency of β_2 integrin subunits results in leukocyte adhesion deficiency and death from infection at an early age.
- Increased adhesion molecule expression may enhance inflammation and play a role in the pathogenesis of rheumatoid arthritis.
- Adhesion molecules can be exploited by viruses, including rhinovirus, that use an adhesion molecule as their receptor to initiate infections in humans.

Study Questions

Choose the ONE best answer.

2.1 Which of the following is a component of the extracellular matrix that is correctly matched with its function?

A. Elastin forms tough glycoprotein fibers that are resistant to shearing forces.

B. Laminin is the principal adhesive glycoprotein in connective tissue.

C. Glycosaminoglycans form a hydrated gel to help resist compressive forces.

D. Fibronectin allow skin and lungs to stretch without tearing.

E. Collagen imparts resilience to tissues such as cartilage.

Correct answer = C. Glycosaminoglycans form a hydrated gel to help resist compressive forces. Collagen is a glycoprotein that helps tissues resist shearing forces. Elastin is a fibrous protein (not glycosylated) that imparts rubbery properties to tissues. Fibronectin is a multifunctional adhesive protein. Glycosaminoglycans form hydrated gels and impart resilience to tissues. Laminin, similar in structure to fibronectin, is an adhesive protein.

2.2 A 78-year-old man is diagnosed with scurvy. His defective collagen results from

   A. A genetic defect that prevents formation of stable collagen triple helices.

   B. Impaired ability to hydroxylate Pro and Lys residues.

   C. Inability to cross-link Lys residues to form desmosine cross-links.

   D. Mutations that replace Gly with other, larger amino acids.

   E. Vitamin C–mediated destruction of tropocollagen molecules.

> Correct answer = B. Impaired ability to hydroxylate Pro and Lys residues. Scurvy is not genetic, but an acquired deficiency of vitamin C. Vitamin C is required for hydroxylation of Pro and Lys residues. Scurvy does not involve mutations, destruction of collagen that has been produced, or ability to form desmosine cross-links. Vitamin C does not mediate degradation of collagen.

2.3 Which of the following properties is unique to cell adhesion mediated by cadherins?

   A. Cadherins are transmembrane proteins; other adhesion molecules are intracellular.

   B. Cadherins mediate cell-to-matrix adhesion, not cell-to-cell adhesions.

   C. Cadherins have homophilic binding, with other cadherins serving as their ligands.

   D. Cadherins mediate bidirectional signaling between the cytoskeleton and the ECM.

   E. Cadherins on one cell bind to the glycosylated ligands on another cell.

> Correct answer = C. Cadherins have homophilic binding, with other cadherins serving as their ligands. The unique characteristic of selectins, not of cadherins, is that they bind to carbohydrate-containing ligands. All adhesion molecules are transmembrane proteins that can mediate cell-to-cell adhesions. Integrins facilitate cell-to-cell and cell-to-ECM adhesions and have inside-out and outside-in signaling.

2.4 A glycoprotein on the surface of a cell is the ligand of a particular adhesion molecule. Therefore, that adhesion molecule most likely belongs to which family?

   A. Cadherins

   B. Collagens

   C. Immunoglobulin superfamily

   D. Integrins

   E. Selectins

> Correct answer = E. Selectins are adhesion molecules that bind to carbohydrate-containing ligands. Cadherins bind to other cadherins. Immunoglobulin superfamily members often have integrins as their ligands. Additionally, integrins can mediate cell to matrix interactions and therefore have ECM components, including fibronectin, as ligands. Collagens are fibrous proteins of the ECM and not a family of adhesion molecules.

2.5 In an individual deficient in α_1-antitrypsin, emphysema may result owing to extensive degradation of

   A. Collagen.

   B. Elastin.

   C. Glycosaminoglycan.

   D. Laminin.

   E. Proteoglycan.

> Correct answer = B. Elastin in lungs can be degraded extensively when α1-antitrypsin, the main inhibitor of elastase, is deficient. Other ECM components, including collagen, glycosaminoglycan, proteoglycan, and laminin are not acted on by elastase. (Note that while the name α1-antitrypsin emphasizes this enzyme inhibitor's role in inhibiting the protease, trypsin, physiologically, it is the main regulator of the proteolytic enzyme elastase. Elastase degrades elastin, especially when not checked by α1-antitrypsin.)

2.6 Which of the following is a component of the extracellular matrix that is correctly matched with its function?

   A. Collagen forms tough protein fibers that are resistant to shearing forces.

   B. Elastin is the principal adhesive glycoprotein in connective tissue.

   C. Fibronectin forms a hydrated gel to allow the ECM to resist compressive forces.

   D. Glycosaminoglycans allow skin and lungs to stretch without tearing.

   E. Laminin imparts resilience to tissues such as cartilage.

> Correct answer = A. Collagen forms tough fibers that impart strength and resistance to shearing forces in tissues that contain it. Elastin is a fibrous protein that imparts rubbery properties to tissues. Fibronectin is a multifunctional adhesive protein. Glycosaminoglycans form hydrated gels and impart resilience to tissues. Laminin, similar in structure to fibronectin, is an adhesive protein.

2.7 A 32-year-old woman with systemic lupus erythematosus (SLE) experiences joint pain and swelling that is accompanied by impaired compression/deformation abilities of her cartilage. Which of the following best describes the impaired extracellular matrix component and the impact it has on cartilage in this patient with SLE?

A. Collagen of SLE cartilage lacks hydroxylysine and is less stable.

B. Elastin of SLE cartilage is unable to stretch so cartilage cannot expand.

C. Excess collagen in SLE cartilage causes joint stiffening.

D. Laminin-mediated adhesion is interrupted, making SLE cartilage more diffuse.

E. Reduced proteoglycan leads to impaired resilience of SLE cartilage.

> Correct answer = E. Reduced levels of proteoglycans in cartilage would impair the resilience of cartilage. Neither collagen, elastin, nor laminin contributes largely to the property known as resilience of cartilage, which results from the hydrate gels formed by the proteoglycans.

2.8 Which of the following properties is unique to cell adhesion mediated by selectins?

A. Selectins are transmembrane proteins; other adhesion molecules are intracellular.

B. Selectins mediate cell-to-matrix adhesion, not cell-to-cell adhesions.

C. Selectins have homophilic binding, with other selectins serving as their ligands.

D. Selectins mediate bidirectional signaling between the cytoskeleton and the ECM.

E. Selectins on one cell bind to the glycosylated ligands on another cell.

> Correct answer = E. The unique characteristic of selectins is that they bind to carbohydrate-containing ligands. All adhesion molecules are transmembrane proteins that can mediate cell-to-cell adhesions. Integrins facilitate cell-to-cell and cell-to-ECM adhesions and have inside-out and outside-in signaling. Cadherins have hemophilic binding and use other cadherins as ligands.

2.9 Fibronectin is the ligand of a particular adhesion molecule. Therefore, that adhesion molecule most likely belongs to which family?

A. Cadherins

B. Collagens

C. Immunoglobulin superfamily

D. Integrins

E. Selectins

> Correct answer = D. Fibronectin is a component of the ECM and integrins are the type of adhesion molecule that can mediate cell-to-ECM adhesions. Cadherins, immunoglobulin superfamily members, and selectins mediate cell-to-cell adhesion only. Collagen is not an adhesion molecule but a fibrous protein within the ECM.

2.10 Therapies that block binding to which of the following types of adhesion molecules may have potential benefit in preventing rhinovirus infection?

A. ICAM-1

B. ICAM-2

C. L-Selectin

D. P-Cadherin

E. VLA-4

> Correct answer = A. Rhinovirus uses ICAM-1 adhesion molecules on endothelial cells of the host's respiratory tract. None of the other adhesion molecules listed are used by rhinovirus in this manner.

Biological Membranes

3

I. OVERVIEW

Membranes form the outer boundary of individual cells and of certain organelles. **Plasma membranes** are the selectively permeable structures of cells that separate interiors from the extracellular environment. Certain molecules are permitted to enter and exit the cell by transport across the plasma membrane.

Plasma membranes are composed of lipids and proteins that form their structure and also facilitate cell function. For example, adhesion and signaling are cellular processes initiated at the plasma membrane. Plasma membranes also serve as attachment points for intracellular cytoskeletal proteins and for components of the extracellular matrix outside of cells.

The basic structure of a biological membrane is a **phospholipid bilayer** (Figure 3.1). Two antiparallel sheets of phospholipids form the membrane that surrounds the inner contents of the cell. The phospholipid membrane layer closest to the cytosol is the **inner leaflet** while the layer closest to the exterior environment is the **outer leaflet**. Cholesterol molecules intercalate or fit between phospholipid molecules. Proteins also associate with the membrane to enable biological functions according to the need of the particular cell, including transport of or response to particular signaling molecules. All these membrane components are important in creating the membrane and establishing a stable, yet dynamic, barrier to maintain the internal environment of the cell while facilitating the biological function of the cell.

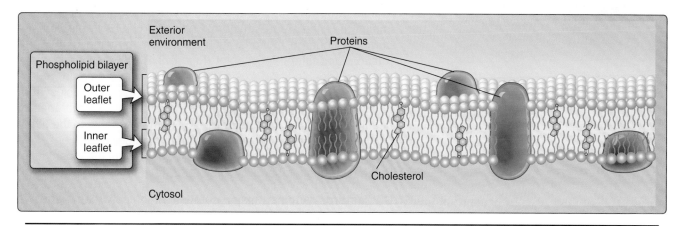

Figure 3.1
Plasma membrane structure.

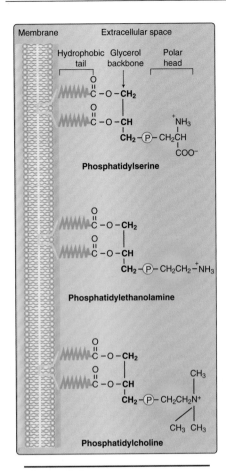

Figure 3.2
Structures of some phospholipids.

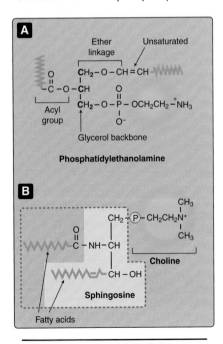

Figure 3.3
Glycerol **(A)** and sphingosine **(B)** backbones in phospholipids.

II. COMPONENTS

All cell membranes, including plasma membranes, organelle membranes, and intracellular vesicles (membrane-enclosed structures inside the plasma membrane), are composed of the same materials. The major components of all biological membranes are lipids and proteins. Several forms of lipids exist to provide structure, support, and function for the membrane. Membrane proteins also play both structural and functional roles.

A. Lipids

In most cell membranes, lipids are the most abundant type of macromolecule present. Plasma and organelle membranes contain between 40% and 80% lipid. These lipids provide both the basic structure and the framework of the membrane and also regulate its function. Three types of lipids are found in cell membranes: phospholipids, cholesterol, and glycolipids.

1. **Phospholipids:** The most abundant of the membrane lipids are the phospholipids. These are polar, ionic compounds that are **amphipathic** in nature. That is, they have both hydrophilic and hydrophobic components. The hydrophilic or polar portion is in the "head group" (Figure 3.2). Within the head group is the phosphate and an alcohol that is attached to it. The alcohol can be serine, ethanolamine, inositol, or choline. Names of phospholipids then include **phosphatidylserine, phosphatidylethanolamine, phosphatidylinositol**, and **phosphatidylcholine.** While all these phospholipids contain a molecule called glycerol, the membrane phospholipid **sphingomyelin** has the alcohol choline in its head group and contains sphingosine instead of glycerol (Figure 3.3).

   The hydrophobic portion of the phospholipid is a long, hydrocarbon fatty acid tail. While the polar head groups of the outer leaflet extend outward toward the environment, the fatty acid tails extend inward to the interior of the phospholipid bilayer. Fatty acids may be saturated, containing the maximum number of hydrogen atoms bound to carbon atoms, or unsaturated with one or more carbon-to-carbon double bonds. (see also *LIR Biochemistry*, Chapter 17). The length of the fatty acid chains and their degree of saturation impact the membrane structure. Fatty acid chains in membranes normally undergo motions such as flexion (bending or flexing), rotation, and lateral movement (Figure 3.4). Whenever a carbon-to-carbon double bond exists, there is a kink in the chain, reducing some types of motion and preventing the fatty acids from packing tightly together. Phospholipids in plasma membranes of healthy cells do not migrate or flip-flop from one leaflet to the other. (However, during the process of programmed cell death, enzymes catalyze the movement of phosphatidylserine from the inner leaflet to the outer leaflet [see also Chapter 23].)

2. **Cholesterol:** Another major component of cell membranes is cholesterol. An amphipathic molecule, cholesterol contains a polar hydroxyl group as well as a hydrophobic steroid ring and attached

hydrocarbon (Figure 3.5). Cholesterol is dispersed throughout cell membranes, intercalating between phospholipids. Its polar hydroxyl group is near the polar head groups of the phospholipids while the steroid ring and hydrocarbon tails of cholesterol are oriented parallel to those of the phospholipids (Figure 3.6). Cholesterol fits into the spaces created by the kinks of the unsaturated fatty acid tails, decreasing the ability of those fatty acids to undergo motion and therefore causing stiffening and strengthening of the membrane.

3. **Glycolipids:** Glycolipids are lipids with attached carbohydrate, and are found in cell membranes in lower concentration than phospholipids and cholesterol. The carbohydrate portion of a glycolipid is always oriented toward the outside of the cell, projecting into the environment. Glycolipids help to form the carbohydrate coat observed on cells and are involved in cell-to-cell interactions. They are a source of blood group antigens and also can act as receptors for toxins including those from cholera and tetanus.

B. Proteins

While lipids form the main structure of the membrane, proteins are largely responsible for many biological functions of the membrane. For example, some membrane proteins function in transport of materials into and out of cells (see Unit III). Others serve as receptors for hormones or growth factors (see Unit IV). The types of proteins within a plasma membrane vary depending on the cell type. However, all membrane proteins are associated with membrane in one of three main ways.

1. **Membrane associations of proteins:** While some proteins span the membrane with structures that cross both bilayers and extend from the environment to the cytoplasm, others are anchored to membrane lipids and still others are only peripherally associated with the cytosolic side of a plasma membrane (Figure 3.7).

   a. **Transmembrane proteins:** Transmembrane proteins are embedded within the lipid bilayer of the membrane and extend from the environment into the cytosol. Some transmembrane proteins contain one transmembrane region while others contain several. Some hormone receptors are proteins with seven distinct membrane-spanning regions (7-pass or 7-loop transmembrane receptors). All transmembrane proteins contain both hydrophilic and hydrophobic components, based on the chemical nature of their amino acid constituents. These proteins are oriented with their hydrophilic portions in contact with the aqueous exterior environment and with the cytosol and their hydrophobic portions in contact with the fatty acid tails of the phospholipids. It is usually the case that proteins cross cellular membranes by adopting a structure containing one or more α **helices** (see *LIR Biochemistry*, Chapters 1 and 2 for a discussion of amino acid and protein structure).

   b. **Lipid-anchored proteins:** Members of the second category of membrane proteins are **lipid-anchored proteins** that are attached covalently to a portion of a lipid without entering the core portion of the bilayer of the membrane.

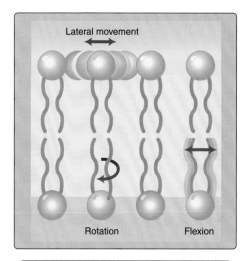

Figure 3.4
Types of motions of membrane phospholipids.

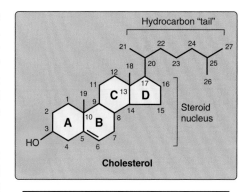

Figure 3.5
Structure of cholesterol.

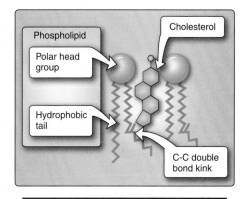

Figure 3.6
Cholesterol and phospholipids in membranes.

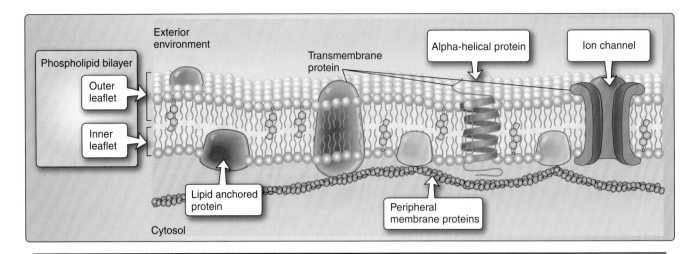

Figure 3.7
Protein associations with membranes.

Both transmembrane and lipid-anchored proteins are **integral membrane proteins** since they can only be removed from a membrane by disrupting the entire membrane structure.

c. **Peripheral membrane proteins:** Proteins in the third category are **peripheral membrane proteins,** which are located on the cytosolic side of the membrane and only attach indirectly to the lipid of the membrane. Such proteins bind to other membrane proteins that are directly attached to the lipids. Cytoskeletal proteins, including those involved in forming the spectrin membrane skeleton of erythrocytes, are examples of peripheral membrane proteins (see Chapter 4).

2. **Membrane protein functions:** Membrane proteins enable cells to function as members of a tissue (Figure 3.8). For example, **cell adhesion molecules** are proteins that extend to the surface of

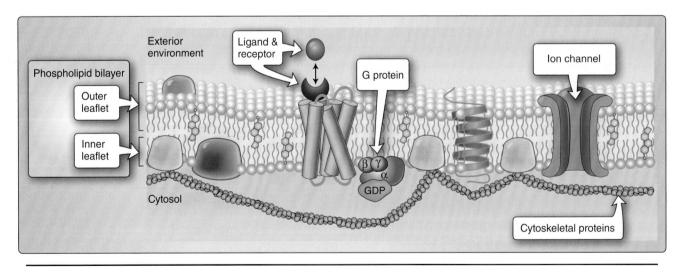

Figure 3.8
Functions of membrane proteins.

cells and facilitate cell-to-cell contact (see Chapter 2). Other membrane proteins function as **ion channels** and **transport proteins** to enable molecules to enter and exit a cell (see Unit III). Membrane proteins that are **ligand receptors** enable cells to respond to hormones and other signaling molecules (see Unit IV). The preceding examples of membrane proteins are of integral, transmembrane proteins whose structures span the bilayer. Lipid-anchored membrane proteins include the **G proteins**, which participate in cell signaling in response to certain ligands (see Chapter 17). Peripheral membrane proteins include **cytoskeletal proteins** that attach to the membrane and regulate its shape and stabilize its structure (see Chapter 4). Some other peripheral membrane proteins are also involved in cell signaling and include enzymes attached to the inner membrane leaflet that are activated after a hormone binds to a protein receptor (see Chapter 17).

III. STRUCTURE

The proteins and lipids of a cellular membrane are arranged in a certain way to form a stable outer structure of the cell. The membrane components, including lipids and proteins, are not fixed rigidly into a particular location. Both can exhibit several types of motions as described previously for phospholipids (see Figure 3.4). Membrane proteins can also move laterally and can rotate. Owing to the composition and dynamic nature of membrane components, the membrane is largely fluid in nature, as opposed to solid or rigid. Despite its fluidity, the membrane structure is very stable and supportive for the cell. The arrangement of the phospholipids provides the basic structure that is then augmented by cholesterol, with functional roles played by proteins.

A. Bilayer arrangement

Membrane phospholipids are oriented with their hydrophobic fatty acid tails facing away from the polar, aqueous fluids of both the cytosol and the environment (such as blood or other cellular fluids including lymph). The hydrophilic portions of the phospholipids are oriented toward the polar environment. Two layers of phospholipids are required to achieve this structure (Figure 3.9). The phospholipids of each layer are found in opposite orientation to each other. While the polar head groups of one layer (outer leaflet) of phospholipids face the exterior, those of the other layer (inner leaflet) face the interior. A nonpolar or hydrophobic central region results where the fatty acid tails of the two layers are in contact with each other.

B. Asymmetry

The fatty acid tails of all the phospholipids are structurally very similar to each other, and the identity of an individual phospholipid molecule is determined by the alcohol within its head group, as mentioned previously (Section II.A.1 above). Some phospholipids are found on the outer leaflet while others are more commonly seen on the inner leaflet. In plasma membranes of most human cells, phosphatidylcholine and sphingomyelin are in the outer leaflet oriented toward the environment, while phosphatidylserine, phosphatidylethanolamine, and phosphatidylinositol are in the inner leaflet oriented toward the cytosol (Figure 3.10).

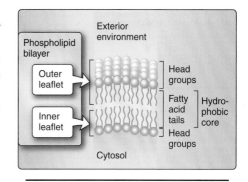

Figure 3.9
Arrangements of membrane phospholipids in a bilayer.

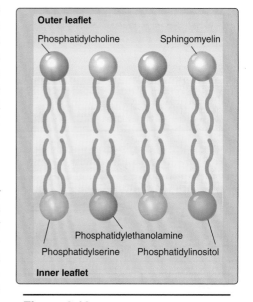

Figure 3.10
Asymmetry of membranes.

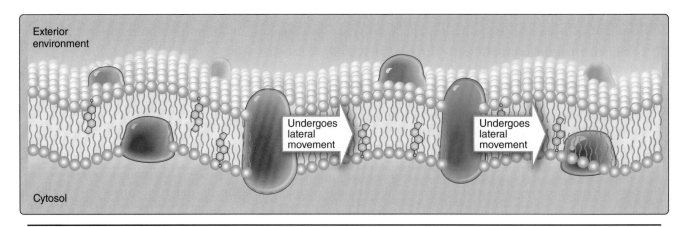

Figure 3.11
Fluid mosaic model.

As also mentioned previously (Section II.A.1 above), during the process of apoptosis or programmed cell death, phosphatidylserine is transferred enzymatically from the inner leaflet to the outer leaflet of the membrane. The presence of phosphatidylserine on the outer leaflet then triggers phagocytic removal of the dying cells, emphasizing further that the maintenance of membrane asymmetry is important for normal cell function.

In addition to an asymmetric distribution of phospholipids between the membrane leaflets, glycolipids are differentially arranged as well and are always on the outer leaflet with their attached carbohydrate projecting away from the cell. Glycoproteins are similarly oriented on the outer leaflet with their carbohydrate portions projecting into the environment. Peripheral membrane proteins are attached only to the inner membrane leaflet, facing the cytoplasm. Therefore, the inner and outer membrane leaflets have different compositions and each has functions distinct from those of the other. Cholesterol, however, can readily flip-flop or move from one leaflet to the other and is distributed on both sides of the membrane bilayer.

C. Fluid mosaic model

For several decades, the membrane model proposed by Singer and Nicholson in 1972 has been used to describe plasma membranes. The membrane is characterized as a fluid, owing to the ability of lipids to diffuse laterally within the plane of the membrane. The overall structure is equated to a flowing sea. And, like a mosaic, membrane proteins are dispersed throughout the membrane. Many of the membrane proteins retain the ability to undergo lateral motion and are likened to icebergs floating within the sea of lipids (Figure 3.11).

D. Lipid rafts

Lipid rafts are specialized sphingolipid and cholesterol-enriched microdomains within cell membranes (Figure 3.12). Functions of lipid rafts include cholesterol transport, endocytosis, and signal transduction. The lipid raft hypothesis assumes that cholesterol combines with glycosphingolipids (phospholipids that have straight acyl chains), to form transient structures that appear as "rafts" floating in the phospholipid sea created by poorly ordered lipids of the surrounding

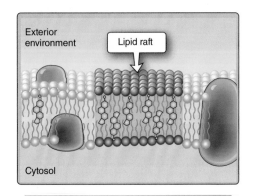

Figure 3.12
Lipid raft.

portions of the membrane. Fatty acid chains of phospholipids within the rafts are extended and more tightly packed. Average sizes, distributions, and lifespan of lipid rafts are not well defined, and the forces that drive their formation are not completely understood. There seem to be strong attractions between sphingolipids and cholesterol and repulsion between phospholipids and the sphingolipids. The repulsive forces likely play a major role in the formation of the rafts. It is difficult to study lipid rafts in living cells and the structures are too small to be observed by light microscopy; however, distinct types of lipid raft structures are described.

Types of lipid rafts include planar, glycosphingolipid-enriched membranes (GEM), and **caveolae.** Planar rafts are continuous with the plane of the plasma membrane and lack any distinctive morphological features. Caveolae, on the other hand, are flask-shaped inward foldings of the plasma membrane that contain the protein **caveolin.** The presence of caveolin causes a local change in morphology of the membrane (Figure 3.13). Caveolae are found in a variety of tissues, particularly in endothelial cells, but are absent from neuronal tissues. Many proteins and lipids are found in high concentrations in caveolae, including arachidonic acid, a fatty acid involved in cell signaling, certain growth factor receptors, integrins, and insulin receptors, among others.

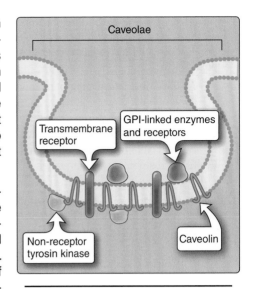

Figure 3.13
Caveolae.

Chapter Summary

- Plasma membranes are selectively permeable outermost structures of eukaryotic cells.
- All biological membranes have the same basic structure.
- Lipids are generally the most abundant type of macromolecule within cell membranes.
- Phospholipids and cholesterol are amphipathic lipids that form the basic structure of cell membranes.
- Proteins associated with membranes may be transmembrane, lipid anchored, or peripheral to the membrane.
- Membrane proteins function as ion channels, transport proteins, ligand receptors, and components of the cytoskeleton.
- The basic membrane structure is that of a phospholipid bilayer.
- An asymmetric distribution of phospholipids results in each side or leaflet of the membrane having distinctive characteristics.
- Lipids and proteins in the membrane are not static but retain the ability to undergo motion within the membrane.
- The fluid mosaic model describes the fluid phospholipid "sea" in which proteins appear to be distributed in a mosaic pattern and to float within the sea of the lipids.
- Membrane microdomains known as lipid rafts are membrane regions enriched in specialized lipids that function in cholesterol transport, endocytosis, and signal transduction.

Study Questions

Choose the ONE best answer.

3.1 Sphingomyelin is identified in a plasma membrane of a healthy, living cell. Which of the following describes its location in that membrane?

   A. The outer membrane leaflet
   B. In a transmembrane arrangement
   C. Intercalated between phospholipids
   D. Anchored to lipids facing the cytosol
   E. Extending into the environment

Correct answer = A. Sphingomyelin is a phospholipid found on the outer membrane lipid. Since phospholipids form the bilayer structure, it cannot be in a transmembrane arrangement or intercalated between phospholipids. Sphingomyelin is a phospholipid, so it is not anchored to lipids and does not extend into the environment.

3.2 A particular molecule of phosphatidylcholine is studied within a plasma membrane of a living cell. Which of the following characteristics is this molecule likely to possess?

  A. A stable, fixed arrangement of its structural components
  B. Continuous flip-flop from one leaflet to the other
  C. Fatty acid tails that undergo flexion
  D. Binding sites to cytoskeletal components
  E. Equal distribution on both sides of the bilayer

Correct answer = C. Phosphatidylcholine is a phospholipid, and fatty acid tails of phospholipids within plasma membranes undergo flexion along with rotation around the phospholipid head and have the ability to move laterally within the plane of the membrane. Membrane phospholipids are not fixed in place but undergo the motions described. However, they do not flip-flop from one membrane leaflet to the other. Phosphatidylcholine is found only on the outer leaflet, not on both sides of the bilayer. It is oriented toward the environment, and not on the inner leaflet toward the cytosol where cytoskeletal components are located.

3.3 The arrangement of a ligand receptor within a plasma membrane is best described as a/an

  A. Peripheral protein
  B. Lipid-anchored protein
  C. Integral membrane protein
  D. Lipid raft
  E. Glycolipid

Correct answer = C. Ligand receptors are proteins that are transmembrane and therefore integral membrane proteins. They are not peripheral membrane proteins or lipid-anchored proteins. Since ligand receptors are proteins, they are not glycolipids and are not lipid rafts.

3.4 A glycoprotein within a plasma membrane has which of the following characteristics?

  A. Attaches to cytoskeletal proteins
  B. Oriented toward the environment
  C. Peripherally attached to the membrane
  D. Found on both the membrane leaflets
  E. Ability to undergo flexion

Correct answer = B. The carbohydrate portions of glycoproteins (and glycolipids) are oriented toward the exterior environment. They are found on the outer leaflet and therefore not on both leaflets or in association with cytoskeletal proteins. Peripheral membrane proteins attach to the inner leaflet. Membrane proteins can move laterally and can rotate but do not possess structures like fatty acid tails that can flex in relation to each other.

3.5 A plasma membrane of a healthy cell is observed to have caveolae. These structures are

  A. Components of phospholipids.
  B. Intercalated between cholesterol molecules.
  C. Composed of disorderly phospholipids.
  D. Regions with high carbohydrate content.
  E. Cholesterol-enriched membrane invaginations.

Correct answer = E. Caveolae are types of lipid rafts which are orderly cholesterol- and sphingolipid-enriched microdomains. Caveolae appear as membrane invaginations or infoldings, caused by the presence of caveolin. They are not components of phospholipids nor composed of disorderly phospholipids. Caveolae do not intercalate between cholesterol molecules but are membrane regions with high cholesterol content.

3.6 Which of the following describes the fluid mosaic model of plasma membranes?

  A. Glycolipid monolayer with glycoproteins on inner leaflet
  B. Cholesterol-enriched microdomains devoid of proteins
  C. Phospholipid bilayer with flowing nature and embedded proteins
  D. Lipid bilayer with components that freely flip-flop between leaflets
  E. Statically arranged layers of phospholipids and cholesterol with proteins on the outside

Correct answer = C. The fluid mosaic model describes the sea of the phospholipid bilayer with embedded proteins likened to icebergs floating within it. A plasma membrane has glycoproteins as only minor components, and they do not form monolayers. Cholesterol-enriched microdomains form lipid rafts, but these are not devoid of proteins, and actually concentrate certain proteins within their structures. Phospholipid bilayers have asymmetrical formation, with certain phospholipids on the inner leaflet and others on the outer. There is no flip-flop between the leaflets in membranes of healthy, living cells. Components of cell membranes have a dynamic nature, as opposed to an unchanging static arrangement. Proteins are found throughout a plasma membrane and are not constrained to one portion.

3.7 A peripheral membrane protein is best described as a protein

A. Embedded within the phospholipid bilayer.
B. That is transmembrane in nature.
C. That traverses the cytosol of a living cell.
D. Enclosed within the membrane infoldings of a caveola.
E. Loosely attached to the inner membrane leaflet.

Correct answer = E. Peripheral membrane proteins are loosely attached to the inner membrane leaflet and are not an integral part of the membrane's structure. Proteins embedded within the phospholipid bilayer and ones that are transmembrane in nature, in contrast, are integral membrane proteins. Proteins that traverse the cytosol of a cell make up the cytoskeleton and are not necessarily by definition peripheral membrane proteins. Proteins within caveolae may have peripheral or integral associations with the membrane, and are not necessarily required to be peripheral in their nature.

3.8 Protein Y is covalently linked to a fatty acid chain of a phospholipid of the inner leaflet of a cell's plasma membrane, but its structure does not extend to the hydrophobic core of the phospholipid bilayer. Therefore Protein Y

A. Has glycosylated regions facing the cytosol.
B. Functions as a ligand receptor.
C. Is a single-pass transmembrane protein.
D. Is oriented towards the cytosolic compartment.
E. Cannot undergo any type of motion in the membrane.

Correct answer = D. A protein attached to the membrane's inner leaflet will be oriented towards the cytosolic component of the cell. Glycoproteins will be found on the outer leaflet, with carbohydrate facing the environment. Ligand receptors are transmembrane, and not attached to the inner leaflet where they would not encounter ligands. A single-pass transmembrane receptor crosses the membrane, from one side to the other, one time. Since this protein does not enter the hydrophobic membrane core, it is not transmembrane. Proteins within membranes undergo motion. Nothing about the description of this particular protein indicates it is unable to rotate, flex, or move laterally within the plane of the membrane.

3.9 A patient with anemia caused by premature destruction of erythrocytes is found to have increased cholesterol content in his erythrocyte membranes. These erythrocyte membranes

A. Exhibit decreased membrane fluidity.
B. Display phosphatidylserine in their outer membrane leaflet.
C. Also contain excess concentrations of collagen.
D. Lack glycoproteins on their surfaces.
E. Have only a single layer of phospholipid forming their structures.

Correct answer = A. Cholesterol content greatly impacts membrane fluidity, with more cholesterol causing decreased fluidity owing to tighter packing of phospholipids when more cholesterol is intercalated between the phospholipids. Phosphatidylserine is found on the inner membrane leaflet of healthy, living cells, and its position is not influenced by the cholesterol content of the membrane. Glycoproteins are found on the outer leaflet of membranes, regardless of cholesterol content of the membranes. Plasma membranes are phospholipid bilayers, not single layers. Cholesterol intercalates between phospholipids, but cholesterol content would not alter the basic membrane bilayer structure.

3.10 When studying membrane glycoproteins, it was determined that removal of a terminal *N*-acetylgalactosamine from a large carbohydrate attached to an erythrocyte membrane protein resulted in those cells being classified as blood type O instead of blood type A. These findings indicate that carbohydrate residues such as *N*-acetylgalactosamine attached to erythrocyte proteins

A. Bury themselves away from the aqueous exterior environment.
B. Face the exterior of the cell and interact with the environment.
C. Flip-flop from the outer to the inner membrane leaflet.
D. Increase the hydrophobicity of lipids in the membranes.
E. Intercalate between phospholipids to increase their packing.

Correct answer = B. Carbohydrate residues face the exterior of the cell and interact with the environment. This is demonstrated in this situation by removal of the carbohydrate causing a reclassification of cells to another blood type. Glycoproteins on the surface of cells are recognized by antibodies to carry out blood typing. Carbohydrates do not bury themselves away from the environment or flip-flop to the inner leaflet. Glycoproteins are not entirely hydrophobic in nature; even if they contain hydrophobic amino acid residues, the carbohydrate portion is hydrophilic. Glycoproteins add to a cell's function more than to membrane structure. It is cholesterol that intercalates between phospholipids to increase their packing.

4 Cytoskeleton

I. OVERVIEW

The **cytoskeleton** is a complex network of protein filaments that creates a supportive scaffolding system within the cell (Figure 4.1). Cytoskeletal proteins are anchored to the plasma membrane and located throughout the **cytoplasm** (interior of the cell), providing a framework in which organelles reside.

The cytoskeleton is not simply a passive internal skeleton but is a dynamic regulatory feature of the cell. **Microtubules** are one type of cytoskeletal protein. They organize the cytoplasm and interact with organelles to induce their movement. In addition to microtubules, **actin filaments** and **intermediate filaments** constitute the cytoskeleton. These components of the cytoskeleton all work together as an integrated network of support within the cytoplasm.

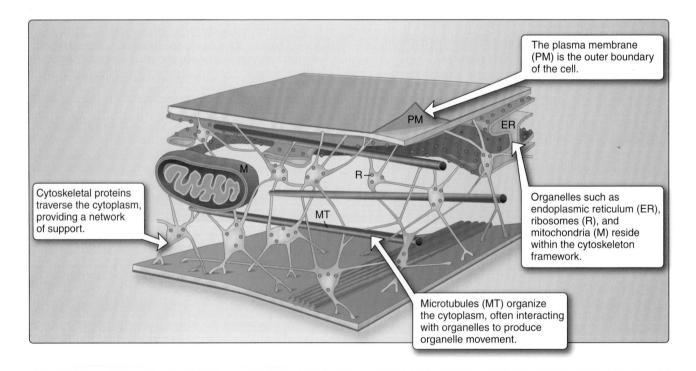

The plasma membrane (PM) is the outer boundary of the cell.

Cytoskeletal proteins traverse the cytoplasm, providing a network of support.

Organelles such as endoplasmic reticulum (ER), ribosomes (R), and mitochondria (M) reside within the cytoskeleton framework.

Microtubules (MT) organize the cytoplasm, often interacting with organelles to produce organelle movement.

Figure 4.1
The cytoskeleton as intracellular scaffolding.

Each type of cytoskeletal filament is formed from a specific association of protein monomer subunits. Actin filaments and microtubules are formed from compact globular protein subunits, while intermediate filaments contain extended fibrous protein subunits. Accessory proteins regulate the filament length, position, and association with organelles and the plasma membrane.

II. ACTIN

First isolated from skeletal muscle, actin was originally believed to be a protein found exclusively in muscle tissues. Four of the six forms of actin are found only in muscle cells, but other forms of actin are found within the cytoplasm of most cell types. Actin is now also believed to be present within the nucleus of most cells. Actin has a diameter of approximately 8 nm and forms structures known as microfilaments. Along with microtubules, actin helps to establish a cytoplasmic protein framework that can be visualized radiating out from the nucleus to the phospholipid bilayer of the plasma membrane (Figure 4.2). Actin is often localized to regions near the plasma membrane referred to as the **cell cortex**.

Actin is important in inducing contraction in muscle cells. Functions of actin in the cytoplasm of nonmuscle cells include regulation of the physical state of the **cytosol** (fluid portion of the cytoplasm without organelles), cell movement, and formation of contractile rings in cell division. Changes in the complexity of structure of actin from the small, globular subunits to elongated polymerized microfilament structures regulate such functions in the cell. In the nucleus, actin is important for stabilizing chromatin and nuclear structure and is believed to be involved in the regulation of gene transcription.

A. Polymerization

Actin microfilaments in the cytoplasm are filamentous or F-actin structures that are polymers formed from individual monomers of G (globular) actin. The polymerization of G-actin subunits into an F-actin molecule is an energy-dependent process.

1. **Overview:** Energy in the form of adenosine triphosphate (ATP) is needed for addition of each G-actin monomer onto a growing F-actin molecule. Each individual G-actin monomer added to a growing F-actin molecule has an ATP molecule bound to it. After the G-actin monomer has polymerized onto the F-actin polymer, ATP is hydrolyzed to adenosine diphosphate (ADP) with the release of inorganic phosphate (P_i) (Figure 4.3). The ADP remains bound to individual G-actin monomers within F-actin polymers. An F-actin filament is composed of two strands of identical G-actin monomers with a structure reminiscent of two strands of beads wound around each other in a regular pattern (Figure 4.4). The ends of F-actin filaments are not identical to each other. The plus (+) end is where the G-actin monomers bound to ATP add to an F-actin filament and the minus (−) end is where the subtraction of ADP-bound G-actin monomers occurs from the F-actin microfilament.

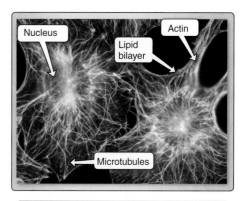

Figure 4.2
Localization of cytoskeletal components.

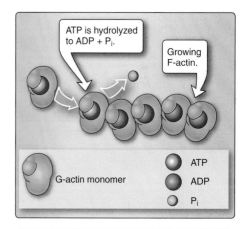

Figure 4.3
ATP hydrolysis in actin polymerization.

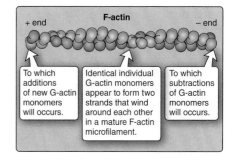

Figure 4.4
Structure of F-actin.

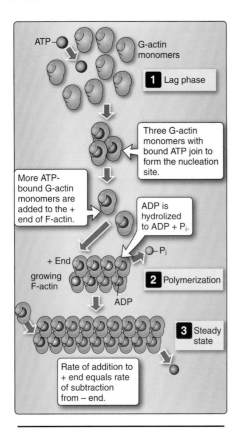

Figure 4.5
Polymerization of actin.

2. **Steps:** An F-actin microfilament structure is generated in three phases: (1) lag, (2) polymerization, and (3) steady state (Figure 4.5). For the lag phase, three G-actin monomers with bound ATP join together to form a nucleation site onto which the growing F-actin filament is built. During the polymerization phase, new G-actin monomers are added to the growing chain, adding preferentially to the + end. ATP is hydrolyzed to ADP + P_i. And, a steady state is reached when addition of G-actin monomers to the + end occurs at the same rate as removal of other G-actin monomers from the − end.

At steady state, the length of the F-actin polymer is maintained. G-actin monomers that leave an F-actin polymer rejoin the cytoplasmic pool of unpolymerized G-actins. The process of addition and subtraction of G-actin monomers is described as "**treadmilling**" because if one observes a particular G-actin monomer during the polymerization process, it will appear to be moving along the length of the F-actin polymer (Figure 4.6) after its initial addition to the polymer. When the next G-actin monomers add onto the polymer after the one of interest, they will seem to push the G-actin being observed further along the polymer. Over time, the G-actin monomer of interest appears to move farther and farther along the F-actin polymer before falling off the other end.

Clinical Application 4.1: Fungal Products and Actin

Some products produced by fungi are known to influence actin polymerization. **Phalloidin** is the toxin from the poisonous mushrooms *Amanita phalloides*. These mushrooms are often called "death caps" or "angels of death" owing to the toxic effects on persons who consume them. Early symptoms are gastrointestinal in nature and are followed by a latency period during which the individual feels better. However, between the 4th and 8th day after the consumption of a mushroom of this type, the person will experience liver and kidney failure and death. Early intervention and detoxification are sometimes helpful, but liver transplantation may still be needed. Phalloidin functions by disrupting the normal function of actin. It binds to F-actin polymers much more tightly than to G-actin monomers and promotes excessive polymerization while preventing filament depolymerization. It also inhibits the hydrolysis of ATP bound to G-actin monomers that have polymerized onto an F-actin polymer. Phalloidin is known to inhibit cell movement due to its inhibition of actin. Phalloidin can be used in the laboratory as an imaging tool to identify actin (see Figure 4.2). The **cytochalasins** are other fungal products that are useful laboratory tools. Because they are known to block polymerization of actin and cause changes in cell morphology, they can be used to inhibit cell movement and cell division and to induce programmed cell death.

B. Actin-binding proteins

Actin-binding proteins regulate the structure of actin within the cell, controlling the polymerization of G-actin monomers, the bundling of microfilaments, and the breakdown into smaller fragments as needed by the cell. Some actin-binding proteins interact with individual G-actin monomers and prevent their polymerization onto an F-actin polymer. Others bind to the assembled microfilaments and cause them either to bundle or cross-link with other microfilament chains or to fragment and disassemble. The complexity of actin-based structures within the cytoplasm regulates certain cell characteristics.

1. **Regulators of the gel-sol of the cytosol:** One characteristic of a cell is the physical nature of its cytosol. It can be described as being either **gel**, a more firm state, or **sol**, a more soluble state. The more structured the actin, the firmer (gel) the cytosol. The less structured (more fragmented) the actin, the more soluble (sol) the cytosol. Actin is continuously treadmilling in both the gel and sol states, contributing to the character of the cytoplasm. In addition, actin-binding proteins regulate the structures of actin and therefore the state of the cytosol (Figure 4.7).

   A pool of G-actin monomers is in equilibrium with F-actin polymers and also with G-actin monomers bound to a sequestering protein that inhibits the ability of G-actin monomers to polymerize. F-actin can be broken into smaller sized fragments by the twisting action of the protein **cofilin** that also prevents further lengthening. F-actin polymers can also be made into more complex structures by the actions of bundling and cross-linking proteins. Those complex structures can be fragmented when needed, by actin-severing proteins such as **gelsolin**, resulting in a more sol state of the cytosol. Calcium is required for function of actin-binding proteins that fragment actin polymers; higher levels of calcium are found in the cytosol where the sol state is developing. For example, the cytosol of a phagocytic cell, such as a macrophage, may need to become less structured with more solubilized cortical actin in order to engulf an invading organism. And, a migrating cell such as a fibroblast moving along a substrate must polymerize the actin at the leading end to pull the cell forward. At the same time, it must solubilize its cortical actin to allow for the flow of contents behind the leading end (Figure 4.8).

2. **Spectrin:** Stability, strength, and support are imparted to cells by actin-binding proteins of the spectrin family. Particularly important in erythrocytes (red blood cells), spectrin has a long, flexible rod-like shape and exists in dimers. On cytosolic face of the plasma membrane, spectrin dimers bind to F-actin filaments associated with phospholipids of the inner leaflet, in an ATP-dependent manner to strengthen and support the erythrocyte membrane. A lattice-like arrangement of actin and spectrin exists with other proteins including ankyrin and protein 4.1 facilitating their interaction.

   The spectrin-actin association is important in maintaining the characteristic biconcave disc shape of erythrocytes that may be important for maximizing the hemoglobin and oxygen carried by each red blood cell (Figure 4.9) (see *LIR Biochemistry*, Chapter 3) and appears to maximize laminar flow in the blood. Erythrocytes

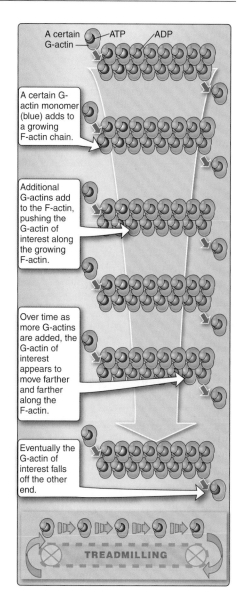

A certain G-actin — ATP   ADP

A certain G-actin monomer (blue) adds to a growing F-actin chain.

Additional G-actins add to the F-actin, pushing the G-actin of interest along the growing F-actin.

Over time as more G-actins are added, the G-actin of interest appears to move farther and farther along the F-actin.

Eventually the G-actin of interest falls off the other end.

TREADMILLING

Figure 4.6
Treadmilling of F-actin.

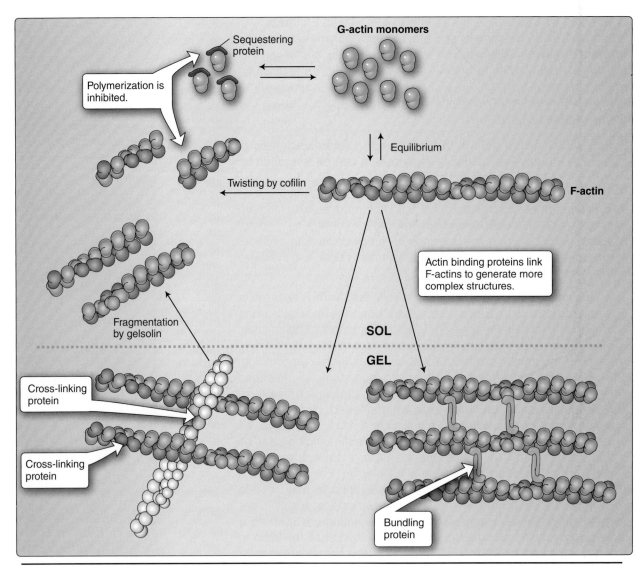

Figure 4.7
Actin and the gel-sol transition.

must also have pliable plasma membranes that can alter and distort their shape when navigating microvasculature. Alterations in spectrin-actin binding facilitate these changes in healthy erythrocytes. Inherited deficiencies that result in absence of spectrin or in presence abnormal spectrin result in **hereditary spherocytosis** (Figure 4.10). As opposed to the usual biconcave disc-shaped erythrocytes, individuals affected with hereditary spherocytosis instead have spherical erythrocytes that are fragile, have less membrane pliability, are susceptible to lysis, and result in hemolytic anemia. Erythrocytes of such individuals are identified via light microscopy by the lack of the central pallor seen in biconcave erythrocytes that have a depression, causing the pale coloration normally observed within erythrocytes.

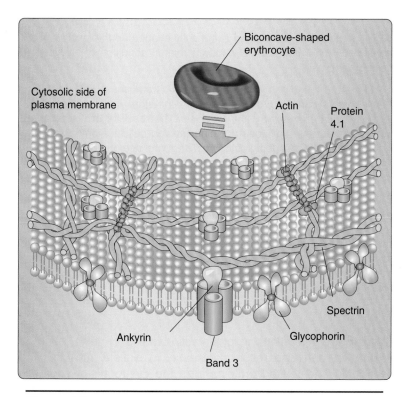

Figure 4.9
The spectrin membrane skeleton in an erythrocyte.

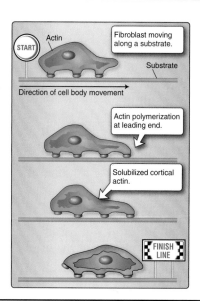

Figure 4.8
Actin polymerization and cell movement.

Clinical Application 4.2: Hereditary Spherocytosis

The prevalence of hereditary spherocytosis is estimated at 1 in 2,000 in individuals of Northern European descent. Inheritance follows an autosomal dominant pattern in most situations, although some instances of autosomal recessive inheritance have been documented. Affected individuals typically experience anemia, accompanied by jaundice and an enlarged spleen. Anemia results when erythrocytes lyse prematurely; jaundice is a consequence of processing of higher than usual levels of hemoglobin from the lysed red blood cells.

Four forms of hereditary spherocytosis are described, based on signs and symptoms: mild, moderate, moderate-to-severe, and severe. The majority of affected individuals have the moderate form of the disease. While those with the mild form sometimes have no symptoms at all, the moderate form typically includes anemia, jaundice, and splenomegaly from childhood. The spleens of affected individuals enlarge owing to the accumulation of misshapen erythrocytes there. Severely affected individuals have these symptoms, in addition to life-threatening anemia that requires frequent blood transfusions. Skeletal abnormalities are often observed in severely affected individuals as well.

Mutations in at least five genes can cause hereditary spherocytosis and related conditions: *ANK1*, which encodes ankyrin-1; *EPB42*, which encodes band 4.2; *SLC4A1*, which encodes an anion exchanger; and *SPTA1* and *SPTB*, which encodes spectrin. Mutations in *ANK1* are the most common cause of hereditary spherocytosis, accounting for more than half. *SPTB*

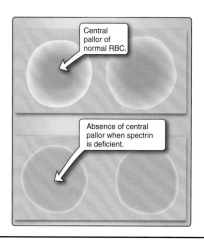

Figure 4.10
Spectrin and erythrocyte shape.

and *SPTA1* mutations also result in spherically shaped erythrocytes, while mutations in *EPB42* and *SLC4A1* cause a related condition of elliptically shaped red blood cells.

The *ANK1* gene is found on the short (p) arm of chromosome 8. At least 55 mutations, some missense and others deletion in nature, in the *ANK1* gene are associated with hereditary spherocytosis. The result of the mutations is an ankyrin-1 protein that does not bind the red cell membrane and is therefore not available for spectrin binding. Therefore, spectrin is completely deficient in the membrane, interfering with erythrocyte membrane shape and flexibility.

SPTB, spectrin beta, erythrocytic, is one member of the spectrin gene family and is located on the long (q) arm of chromosome 14. Various mutations in this gene are associated with autosomal dominant spherocytosis or with autosomal dominant elliptocytosis. Altered spectrin produced from the mutant gene cannot associate properly with band 4.1 and actin in the erythrocyte membrane to create and maintain a flexible, biconcave disc morphology.

3. **Dystrophin:** Alterations in other actin-binding proteins, such as dystrophin, can also result in disease. In skeletal muscle cells, dystrophin and related proteins form the dystrophin-glycoprotein complex that links actin to the basal lamina (see Chapter 2). This association between dystrophin and actin provides tensile strength to muscle fibers and also appears to act as a framework for signaling molecules. Defects in dystrophin result in muscular dystrophy (MD), a group of genetic disorders whose major symptom is muscle wasting.

Clinical Application 4.3: Forms of Muscular Dystrophy

When the dystrophin protein is absent or is present in a nonfunctional form, degeneration of muscle tissue results. Muscle wasting then follows when the ability to regenerate the muscle is exhausted. **Duchenne MD** (DMD) and **Becker MD** are both caused by mutations in the dystrophin (dys) gene, a large (2.6 Mbp) gene with 97 exons. Additionally, both are inherited as X-linked recessive traits, with male individuals expressing the disease when their only X chromosome carries the mutated *dys*. These forms of MD differ in age of onset and in severity. Symptoms of DMD are noticeable in early childhood and quickly become debilitating. Becker MD is characterized by slowly progressive muscle weakness in the pelvis and legs. It has a later age of onset than DMD and less severe symptoms. Several large deletions in the *dys* gene are found in individuals with Becker MD. These deletions, however, do not result in a frameshift (see Chapter 9). Instead, internal portions of the gene are missing, but transcription and translation of the remainder of the gene occur. A partially functional dystrophin protein is therefore synthesized, causing less severe muscle wasting. In contrast, in DMD, several smaller deletions are found in the *dys* gene. In DMD, however, there is a frameshift, resulting in early termination of the protein (see Chapter 9). No functional dystrophin is produced and more severe muscle degeneration occurs.

C. Contractile functions in nonmuscle cells

While actin polymerization helps to propel a cell forward (see Figure 4.8), contraction is needed to pull the plasma membrane of the lagging end away from the substrate to enable forward progress. In fact, contraction or tightening and shortening to produce a pulling force is needed to maintain cell structure and to carry out normal cellular functions. Actin participates in such contractions owing to the effects of an ATP-hydrolyzing motor protein of the **myosin** family.

1. **Myosin:** As is true for actin, myosin was first discovered in muscle but is found in all cell types. Actin-myosin interactions in muscle are well studied (see *LIR Physiology*, Chapter 12). It is believed that similar mechanisms function to produce contractions in nonmuscle cells. Myosin molecules have a head domain that interacts with F-actin and a tail containing an ATP-binding site. They hydrolyze ATP when they bind to actin. Myosin's interactions with actin are cyclic. Myosin binds to actin, detaches, and then binds again. The myosin tail can also attach to cellular structures and pull them along the actin filament. Several forms of myosin exist. In nonmuscle cells, myosin II is important for many examples of contraction as it slides actin filaments over each other to mediate local contractions (Figure 4.11A). Myosins I and V are involved in moving cellular cargo along the tracks provided by F-actin.

2. **Structural and functional needs for contraction:** Providing stability to cell structure is an important reason for actin-myosin–based contractions to occur. Myosin II interacts with F-actin in the cell cortex to stiffen it and help prevent deformation of the plasma membrane. In addition, actin bundles that encircle the inner portion of epithelial cells form a tension cable, known as a circumferential belt that can regulate cell shape. This type of contraction is important in wound healing because the gap of the wound can be sealed via contraction of the existing cells. Cell division in all cell types also depends on contraction. Actin-myosin structures are important during cytokinesis, or the cytoplasmic division following nuclear division of mitosis (see Chapter 20). Here, an actin-myosin–based structure, called a **contractile ring**, is formed. The diameter of the contractile ring progressively decreases, deepening the cleavage furrow in order to pinch the cell into two daughter cells. The ATP hydrolysis by myosin attached to actin causes pulling on the actin and pinching and separation of the membrane (Figure 4.11B).

III. INTERMEDIATE FILAMENTS

With a 10-nm diameter, the appropriately named intermediate filaments are larger than actin microfilaments and smaller than microtubules. Most intermediate filaments are located in the cytosol between the nuclear envelope and the plasma membrane. They provide structural stability to the cytoplasm, somewhat reminiscent of the way that steel rods can reinforce concrete. Some other intermediate filaments, the nuclear lamins, provide strength and support to the nucleus. There are six categories of intermediate filaments, grouped by their location. All have structural characteristics in common (Table 4.1).

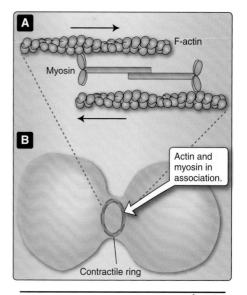

Figure 4.11
Contractile functions of actin in nonmuscle cells. **A.** Myosin II slides actin filaments over each other to mediate local contractions. **B.** A contractile ring pinches a dividing cell into two daughter cells.

Table 4.1: Types of Intermediate Filaments

Type	Names	Functions
I and II	Acidic (I) and basic (II) keratins	Form complex network from nucleus to plasma membrane in epithelial cells
III	Desmin, vimentin	Support and structure
IV	Neurofilaments, synemin, syncoilin	Protect from mechanical stress and maintain structural integrity in various cell types
V	Nuclear lamina	Structural role in the nucleus of all cells
VI	Nestin	Expressed mainly in the nerve cells and is implicated in their growth

A. Structure

Intermediate filaments are formed by α-helical rod-like protein sub-units that have globular domains at their amino-terminal and carboxy-terminal ends. Two rod-like subunits combine to form dimers, known as **coiled coils** (Figure 4.12). One coiled-coil dimer self-associates with another coiled-coil dimer, in a staggered pattern, to form a tetra-mer. Because the carboxy-terminal end of one coiled coil is in close proximity to the amino-terminal end of the other coiled coil, tetramers have an antiparallel orientation. Tetramers attach to each other in a side-to-side, staggered array of eight tetramers that wind together

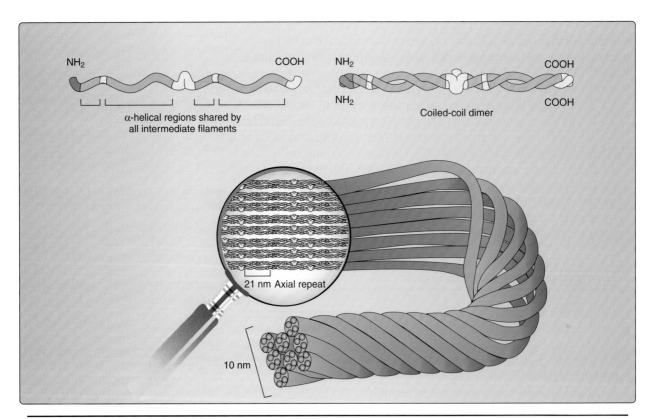

Figure 4.12
Structure of intermediate filaments.

to form the rope-like structure of the mature intermediate filament. No energy is required for the assembly of intermediate filaments. Subunits of intermediate filaments are always found incorporated into stable structures. They lack polarity and therefore do not have a + or a – end. The amino- and carboxy-terminal regions are specific to each class of intermediate filaments.

B. Types of intermediate filaments

Six categories of intermediate filaments have been defined, based on their similarities in structure and on the location in which they function. Types I and II are the **keratins**, with the first being acidic keratins and the second basic. Acidic and basic keratins bind to each other to form functional keratins found in epithelial cells. Type III contains four members, including **vimentin**, the most widely distributed intermediate filament protein. Type IV is found in neurons and type V is nuclear lamina found in all nucleated cells to provide structural support in the nucleus.

IV. MICROTUBULES

Microtubules are the last type of predominant structure observed in the cytoskeleton. They are involved in chromosomal movements during nuclear divisions (mitosis and meiosis), in formation of cilia and flagella in certain cell types, and in intracellular transport. Similarities are seen between actin and microtubules with regard to an energy requirement for their assembly and their ability to undergo structural changes according to the needs of the cell.

To coordinate and regulate microtubules based on cellular requirements, cells contain a structure called a **centrosome**. Present on one side of the nucleus when the cell is not in mitosis, centrosomes organize microtubules by regulating their number, location, and cytoplasmic orientation. Microtubules are seen to radiate outward from the centrosome with some growing in length, while others are shrinking or disappearing all together. Although microtubules function independently from other microtubules, each one has the same basic structure.

A. Structure

The structure of a microtubule resembles a hollow cylindrical tube. Its basic structural component is a protein, composed of an α and a β **tubulin** dimer. Linear chains of $\alpha\beta$ tubulin heterodimers self-assemble into structures called **protofilaments**. A ring of 13 tubulin molecules, which is 24 nm in diameter and embedded in the centrosome, forms the nucleation site onto which the microtubule is built. The β tubulin end of the $\alpha\beta$ tubulin heterodimer appears to be oriented away from the centrosome. Thirteen protofilaments then form the outer wall of the cylindrical microtubule structure, which will vary greatly in length depending on the polymerization status and the function of the microtubule (Figure 4.13).

B. Assembly

The polymerization or assembly of individual microtubules is a complex process because microtubules continuously switch back

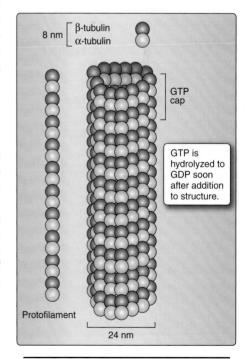

8 nm
β-tubulin
α-tubulin

GTP cap

GTP is hydrolyzed to GDP soon after addition to structure.

Protofilament

24 nm

Figure 4.13
Structure of microtubules.

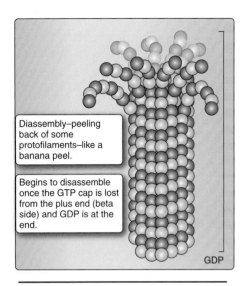

Figure 4.14
Microtubule disassembly.

and forth between growing and shrinking phases. Because of their ever-changing growth status, microtubules are described as having "**dynamic instability**." In the assembly process, tubulin heterodimers bound to guanosine triphosphate (**GTP**) are able to interact with other GTP-bound tubulin heterodimers to form protofilaments. Soon after their polymerization onto a growing microtubule, the GTP on the tubulin is hydrolyzed to guanosine diphosphate (GDP), in a manner similar to the ATP-to-ADP hydrolysis seen in actin protofilaments. A **GTP cap** is described on the + or leading end (opposite end of the one anchored in the centrosome) of the microtubule, representing the newly added tubulin heterodimers on which the GTP has not been hydrolyzed to GDP. The microtubule continues to extend its length, in search of cellular structures, such as organelles or chromosomes, to which it can bind. Growth of the microtubule continues until it either binds to such a structure or until it loses a critical mass of GTP-bound tubulins from the leading end.

C. Disassembly

When addition of GTP-bound tubulins to the protofilaments slows down, hydrolysis of GTP to GDP catches up and the GTP cap is lost. New tubulin heterodimers cannot bind to GDP-containing tubulin heterodimers. In addition to stopping microtubule growth, lack of a GTP cap destabilizes the structure. Individual protofilaments peel back and curve away from the center of the cylindrical tube (Figure 4.14). Tubulin heterodimers containing GDP dissociate from the protofilament, and the microtubule is disassembled very rapidly. It may disappear completely, or it may start growing again if GTP replaces GDP bound to tubulin heterodimers. New microtubules quickly form to replace those that have been disassembled.

D. Functions

While new microtubules being generated radiate out from their centrosome, other microtubules are disassembling, or depolymerizing. However, if a growing microtubule can bind stably to a cellular structure, depolymerization will be prevented. Proteins can bind to and stabilize microtubules to inhibit their disassembly. Stabilized microtubules can then assist in organizing the cytosol and facilitating chromosomal movements and transport along the microtubule network.

1. **Chromosomal movements:** Microtubules pull and push chromosomes in dividing cells to enable the segregation of genetic material into newly formed daughter cells (see Chapter 20). In mitosis, where one parent cell is duplicated to generate two identical daughter cells, the nuclear envelope surrounding the nucleus must break down. Cytoplasmic microtubules disassemble. Then, microtubules reassemble into an organized structure called the **mitotic spindle** (Figure 4.15). This structure is stable but dynamic, as there is a continual exchange between unpolymerized tubulin heterodimers and tubulin molecules polymerized onto microtubules. Because the barrier of the nuclear envelope is gone, the microtubules can gain access to the chromosomes. Microtubules are stabilized by this binding and their assembly and disassembly cease in order for them to carry out this function.

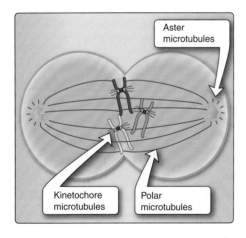

Figure 4.15
Mitotic spindle.

Microtubules align the chromosomes at metaphase, pull them apart at anaphase, and move them toward opposite poles of the cell in telophase. Microtubules within a mitotic spindle have specified roles. Polar microtubules push the spindle apart, while kinetochore microtubules attach to the kinetochore structures of the duplicated chromosomes (see Chapter 20). Astral microtubules that radiate out from the centrosomes are believed to position the spindle.

2. **Formation of cilia and flagella:** Some microtubules form stable structures of cilia and flagella, according to the specific needs of the cell. **Cilia** (singular "cilium") are important in movement of fluids, such as mucus, across epithelial cells of the respiratory tract. They move in a cyclic manner, taking strokes through the fluid. **Flagella** (singular "flagellum") are generally longer than cilia and are important in moving an entire cell, such as sperm, through fluids. Both cilia and flagella have structures dependent upon microtubules. Nine specialized pairs of microtubules form a ring and surround two additional microtubules (Figure 4.16). Microtubules within these structures bend and slide against each other, producing movement. **Dynein** is a microtubule-binding protein that generates the sliding forces between microtubules in cilia, catalyzing their movement.

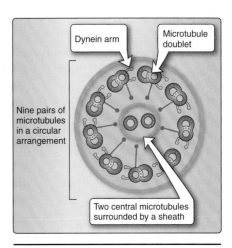

Figure 4.16
9 + 2 arrangement of microtubules in cilia and flagella.

Mitotic spindle poisons

Certain compounds can inhibit or halt cell division by interfering with microtubules. **Colchicine**, for example, binds to the unpolymerized tubulin molecules and prevents their polymerization onto a growing microtubule. If colchicine is given to cells undergoing division, their mitotic spindle breaks down. Abnormally dividing cells are unable to survive. Thus, compounds related to colchicines, such as vinca alkaloids vinblastine and vincristine, are often used to treat uncontrolled cell growth in cancer (see Chapter 23). Another anticancer drug, **Taxol**, also binds to tubulin. However, it preferentially binds to tubulin within assembled microtubules and prevents disassembly. The inability of microtubules to undergo structural changes within the mitotic spindle leads to an arrest of the dividing cells in mitosis.

E. Microtubule motor proteins

Some other types of movement within cells are also dependent upon microtubules. For example, organelles as well as vesicles can be observed to travel along microtubules within cells. **Dynein** and another microtubule-binding protein, **kinesin**, facilitate movement of intracellular cargo that can include membrane-bound organelles and transport vesicles. Dynein and kinesin are actually families of proteins called microtubule motor proteins that have ATP-binding heads and tails that bind stably with their intracellular cargo. Dyneins move along microtubules toward the centrosome (toward the − end of the microtubules), while kinesins travel along microtubules away from the centrosome (toward the + end of the microtubules). Both types of proteins hydrolyze ATP to catalyze their own movement along microtubules while pulling their cargo along the network provided by the microtubules (Figure 4.17).

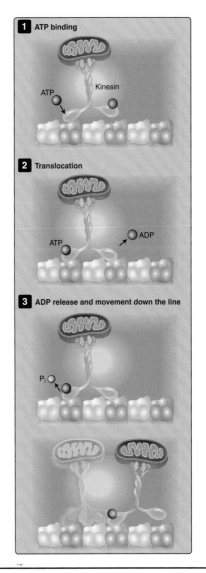

Figure 4.17
Microtubule motors and organelle transport.

Chapter Summary

- The **cytoskeleton** is a complex network of protein filaments found throughout the interior of cells.
- The three principal types of protein filaments of the cytoskeleton are **actin** filaments, **microtubules**, and **intermediate filaments**.
- Accessory proteins bind to and regulate the function of cytoskeletal proteins.

Actin:

- Functions of actin in nonmuscle cells include regulation of the physical state of the **cytosol**, cell movement, and formation of contractile rings in cell division.
- Actin polymerization occurs by the addition of G-actin monomers onto F-actin polymers, in an ATP-dependent manner.
- **Treadmilling** is the dynamic process of addition of a new G-actin monomer to a growing chain followed by its displacement along the F-actin polymer and its removal from the chain as new G-actin monomers join the chain in front of it.

Intermediate filaments are stable, rope-like cytoskeletal structures that provide strength and support.

Microtubules:

- Microtubules are involved in chromosomal movements during nuclear division, in formation of cilia and flagella, and in intracellular transport.
- Composed of **tubulin** heterodimers, microtubules have a **dynamic instability** and continue to assemble and disassemble according to the needs of the cell.

Study Questions

Choose the ONE best answer.

4.1 A cytoskeletal component requires ATP for its polymerization and contains subunits that are observed to undergo treadmilling. That cytoskeletal component is a/an

    A. Actin filament.
    B. Intermediate filament.
    C. Keratin.
    D. Microtubule.
    E. Tubulin.

> Correct answer = A. Actin filaments require ATP for polymerization, and subunits undergo treadmilling as they make their way through an F-actin polymer. Intermediate filaments have stable structures and do not undergo dynamic assembly or disassembly processes. Keratin is a type of intermediate filament. Microtubules are composed of tubulin heterodimers. They require GTP for their polymerization. Additions and subtractions of tubulins from microtubules occur from the same end of the structure.

4.2 A 4-year-old boy presents in the clinic with signs and symptoms of muscular dystrophy. Testing reveals a small deletion in the *dys* gene and no functional dystrophin protein. Owing to this defect, which of the following cytoskeletal components is unable to stably bind to the basal lamina in skeletal muscle cells?

    A. Actin filaments
    B. Intermediate filaments
    C. Microtubules
    D. Myosin
    E. Vimentin

> Correct answer = A. Dystrophin facilitates actin binding to the basal lamina of the skeletal muscle cells. Neither intermediate filaments nor microtubules bind in this manner. Myosin is another actin-binding protein that facilitates contraction of F-actin. Vimentin is a type of intermediate filament.

4.3 A microtubule is observed to disassemble quickly after a period of rapid growth. Which of the following most likely occurred to this particular microtubule to stimulate its breakdown?

A. Binding by dynein
B. Formation of its ATP cap
C. Loss of its GTP cap
D. Severing by gelsolin
E. Twisting by cofilin

Correct answer = C. Loss of the GTP cap results in rapid disassembly of microtubules. Dynein is a microtubule motor protein that enables microtubules to facilitate movement of cilia and flagella and of intracellular cargo. ATP is not part of a microtubule structure. Gelsolin and cofilin are actin-binding proteins that stimulate breakdown of complex actin structures.

4.4 Rapidly dividing cells are grown in the laboratory in the presence of the drug, Taxol. Which of the following effects may be observed in the cytoskeleton of these cells?

A. Continued growth of microtubules
B. Conversion from gel to sol state of cytosol
C. Halt in treadmilling of F-actin
D. Intermediate filament disassembly
E. Tubulin depolymerization

Correct answer = A. In the presence of Taxol, the disassembly of microtubules is prevented, and microtubules will continue to grow in length. The status of cortical actin regulates the gel-to-sol transition in cells. Taxol binds to tubulins within microtubules and does not affect actin or intermediate filaments. Depolymerization of microtubules is prevented by Taxol binding.

4.5 A vesicle within a cell must be transported to another region of the cell along the microtubules. Which of the following proteins may be involved in catalyzing this transport?

A. Dystrophin
B. Kinesin
C. Myosin
D. Spectrin
E. Vimentin

Correct answer = B. Kinesin and dynein are families of microtubule motor proteins that facilitate intracellular transport along microtubules. Dystrophin, myosin, and spectrin are actin-binding proteins. Vimentin is a type of intermediate filament.

4.6 Actin polymerization may function to control or regulate

A. Changes in the physical state of the cytosol.
B. Chromosomal movement during cell division.
C. Provide rigid structural stability to the cytoplasm.
D. Resilience of tissue such as cartilage.
E. Strength within connective tissues.

Correct answer = A. Actin polymerization controls the physical state of the cytosol and the transition from gel to sol. Microtubules, composed of tubulin, regulate chromosomal movements in cell division. Intermediate filaments provide structural stability to the cytoplasm. Unlike actin and microtubules that have dynamic changes in structure, intermediate filaments are longer-lived more permanent structures and may be likened to rigid supports. Resilience is a property attributed to proteoglycans in the extracellular matrix, not to cytoskeletal components such as actin. Collagen and elastin lend strength to the ECM in connective tissue.

4.7 An 8-month-old female patient presents with jaundice and splenomegaly. Her hemoglobin is below reference range and her serum lactate dehydrogenase is markedly elevated, indicating cell lysis. A peripheral blood smear reveals small, globular erythrocytes lacking the central pallor. These findings are best explained by erythrocyte deficiency of

A. Actin.
B. Collagen.
C. Elastin.
D. Glycosaminoglycan.
E. Spectrin.

Correct answer = E. Spectrin deficiency within erythrocyte membranes causes membrane changes and spherically shaped erythrocytes that lyse readily. Spectrin is an actin-binding protein, but actin is not deficient in cells of individuals with hereditary spherocytosis. Collage, elastin, and glycosaminoglycan are all components of the ECM and not of individual cells.

4.8 A 17-year-old male is evaluated for slowly progressive muscle weakness in his pelvis and legs. Several large deletions are found in a gene that encodes an actin-binding protein, resulting in a partially functional protein. This patient is most likely affected with

A. Becker muscular dystrophy.
B. Ehlers-Danlos syndrome.
C. Hereditary spherocytosis.
D. Marfan syndrome.
E. Pemphigus vulgaris.

Correct answer = A. The actin-binding protein described in dystrophin. Ehlers-Danlos syndrome is caused by inherited defects in fibrillary collagen with excessively stretchy skin and hyperextendable joints characteristic of most forms. Hereditary spherocytosis is caused by lack of erythrocyte spectrin, preventing formation of the usual biconcave disc membrane conformation of erythrocytes. Marfan syndrome involves mutation in fibrillin-1, a protein essential for maintenance of elastic fibers, with pathology observed in eyes, skeletal system, and the aorta. Pemphigus vulgaris, characterized by mouth sores, occurs from a disruption of cadherin-mediated cell adhesions.

4.9 A 28-year-old male consumes mushrooms with are poisonous *Amanita phalloides*. The phalloidin toxin from these mushrooms disrupts normal cell function by binding tightly to

A. Acidic keratins.
B. Cadherins.
C. Desmosomes.
D. F-actin polymers.
E. Tubulin heterodimers.

Correct answer = D. Phalloidin is a fungal toxin that binds to and disrupts the function of actin. It does not interfere with structure of keratins, with cell adhesion of cadherins or desmosomes and does not affect tubulin heterodimers or the microtubules they form.

4.10 A 24-year-old female patient diagnosed with Hodgkin disease is treated with combination chemotherapy. Her drug regimen includes Velban, known to inhibit microtubule formation. Therefore, which of the following processes will be altered/impaired by Velban?

A. Formation of the mitotic spindle with arrest of malignant cells in mitosis
B. Production of filamentous actin from G-actin monomers within malignant cells
C. Stabilization of malignant cell membranes and protection from stretching forces
D. Transformation of cytosol in malignant cells from the gel to the sol states
E. Treadmilling of filamentous tubulin monomers in an ATP-dependent process

Correct answer = A. Inhibitors of microtubule formation such as Velban will prevent mitotic spindle formation with arrest of malignant cells in mitosis. Actin is not affected so formation of F-actin will continue and transformation from gel to sol state, regulated by actin, will also continue in the presence of the drug. Tubulin monomers are globular heterodimers, not filamentous. And GTP, not ATP, is used in formation of microtubules. Treadmilling is usually described for actin, not for microtubules.

Organelles

5

I. OVERVIEW

Organelles are complex intracellular structures where processes necessary for eukaryotic cellular life occur. Most organelles are enclosed by membranes composed of the same components as plasma membranes that form the outer boundaries of cells (see Chapter 3). Together with the **cytosol** (gel-like intracellular contents), the organelles help to form the **cytoplasm**, composed of all materials contained within the boundaries of the plasma membrane. Organelles do not float freely within the cytosol but are interconnected and joined by the framework established by proteins of the cytoskeleton (see Chapter 4).

Each organelle carries out a specific function although the activities of organelles can also sometimes be united. Cooperation between organelles is necessary for the expression of genes encoded by nuclear DNA as proteins that function in various intracellular and extracellular locations. Organelles in this group include the **nucleus, ribosomes,** the **endoplasmic reticulum (ER),** and the **Golgi complex**. Members of this collection of organelles have a characteristic arrangement within the cell and their proximity to each other allows them to carry out their function in protein processing (Figure 5.1). Starting with the nucleus and working outward toward the plasma membrane, the ER with attached ribosomes is found next, followed by the Golgi complex, which is in close proximity to the plasma membrane.

Other organelles are found in various locations within the cytoplasm and have functions distinct from but equally important to those involved in protein processing. The main function of mitochondria is to harvest energy to power cells' metabolic processes. Some other organelles are involved in digestion and detoxification. **Lysosomes** contain potent enzymes that break down macromolecules at the end of their life span while **peroxisomes** have functions that include detoxification of peroxides that would otherwise damage the cell.

II. ORGANELLES IN PROTEIN PROCESSING

The processes involved in the expression of DNA as functional proteins require cooperative actions by the nucleus, ribosomes, the ER, and the Golgi complex. The details of protein processing and trafficking or

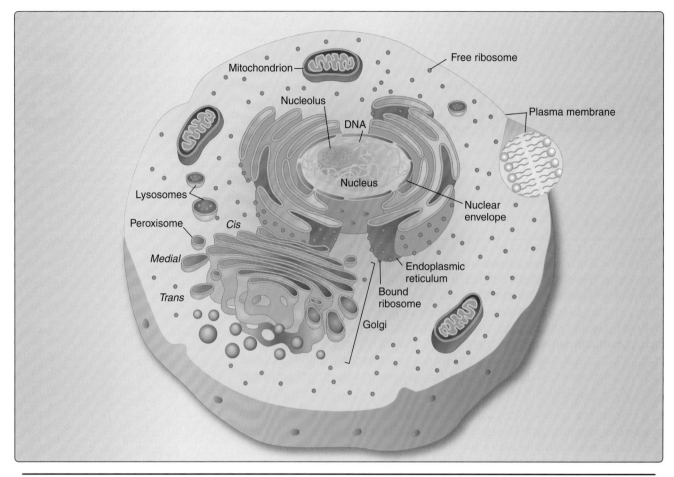

Figure 5.1
Diagram of a eukaryotic cell showing characteristic features and arrangements of organelles.

movement between organelles will be discussed in Chapter 11. The structures and functions of each of these organelles are our present focus.

A. Nucleus

All eukaryotic cells, except mature erythrocytes (red blood cells), contain a nucleus (plural = nuclei) where the cell's genomic DNA resides. In cells that are not actively dividing, the DNA is contained within chromosomes (see also Chapter 22). Every normal human cell contains 23 pairs of chromosomes within the nucleus of every cell. The outermost structure of the nucleus is the **nuclear envelope** (Figure 5.2). This is a double-layered phospholipid membrane with **nuclear pores** to permit transfer of materials between the nucleus and the cytosol. The interior of the nucleus contains the **nucleoplasm**, the fluid in which the chromosomes are found. This is organized by the **nuclear lamina**, the protein scaffolding of the nucleoplasm that is composed mainly of intermediate filaments (see also Chapter 4). The nuclear lamina forms associations between the DNA and the inner nuclear membrane. A prominent structure within the nucleus is a suborganelle called the **nucleolus**, the site of **ribosome** production.

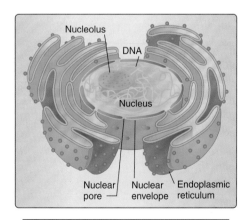

Figure 5.2
Structure of a cell's nucleus.

B. Ribosomes

Ribosomes are the cellular machinery for protein synthesis (see also Chapter 9) and are composed of proteins and ribosomal RNA (rRNA) (Figure 5.3) with approximately 40% being protein and 60% rRNA. Ribosomes are found within the cytosol either free or else bound to the ER. A complete ribosome has two subunits, one large and the other small. The large subunit contains three rRNA molecules and close to 50 proteins while the small subunit has one rRNA and approximately 30 proteins. Ribosomes assemble when needed for translation (protein synthesis) of protein from messenger RNA. Ribosomal subunits disassemble after completing the translation of a particular mRNA.

C. Endoplasmic reticulum

Appearing like a series of interconnected, flattened tubes, the **ER** is often observed to surround the nucleus (Figure 5.4). The outer layer of the nuclear envelope is actually contiguous with the ER. In muscle cells, this organelle is known as the sarcoplasmic reticulum. The ER forms a maze of membrane-enclosed, interconnected spaces that constitute the ER **lumen**, which sometimes expand into sacs or **cisternae**. Regions of ER where ribosomes are bound to the outer membrane are called **rough endoplasmic reticulum** (rough ER or rER). Bound ribosomes and the associated ER are involved in the production and modification of proteins that will be inserted into the plasma membrane; function within lysosomes, the Golgi complex, or the ER; or else will be secreted outside the cell (see also Chapter 11). **Smooth endoplasmic reticulum** (sER) refers to the regions of ER without attached ribosomes. Both rER and sER function in the glycosylation (addition of carbohydrate) of proteins and in the synthesis of lipids.

D. Golgi complex

Working outward from the nucleus and the ER, the next organelle encountered is the Golgi complex. This organelle appears as flat, stacked, membranous sacs (Figure 5.5). Three regions are described within the Golgi complex: the **cis**, which is closest to the ER, the **medial** in the center, and the **trans** Golgi, which is nearest the plasma membrane. Each region is responsible for performing distinct modifications, such as glycosylations, phosphorylations (addition of phosphate), or proteolysis (enzyme-mediated breakdown of protein), to the newly synthesized polypeptides being converted into mature, functional proteins. The trans Golgi network sorts and packages the newly synthesized and modified proteins into distinct regions within the trans Golgi. These regions bud off from the main body of the Golgi complex and form structures called transport vesicles. Movement of these new proteins toward their final cellular or extracellular destination is facilitated in this manner.

III. MITOCHONDRIA

Mitochondria (singular = mitochondrion) are complex organelles that have several important functions in eukaryotic cells. Their unique membranes are used to generate ATP, greatly increasing the energy yield from the breakdown of carbohydrates and lipids. Mitochondria can self-replicate

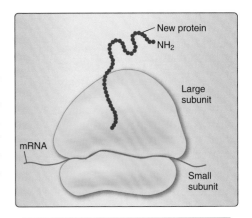

Figure 5.3
A ribosome synthesizing protein from mRNA.

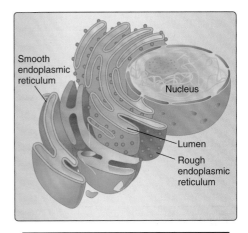

Figure 5.4
ER forming a contiguous membrane structure with the nucleus.

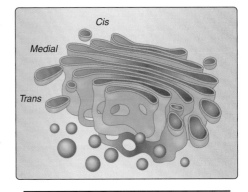

Figure 5.5
Golgi complex.

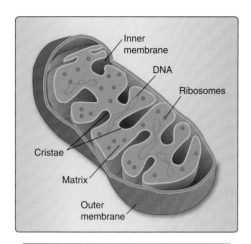

Figure 5.6
A mitochondrion.

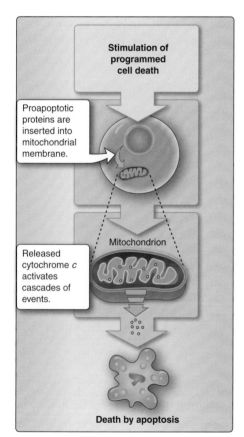

Figure 5.7
Mitochondria in apoptosis.

(reproduce themselves autonomously), and they also contain their own DNA. Owing to these properties, mitochondria are believed to have their origins as single-celled prokaryotic organisms. The very survival of individual eukaryotic cells depends on the integrity of their mitochondria. Programmed cell death or apoptosis occurs when pores are formed in the mitochondrial membrane allowing for the release of proteins that facilitate the apoptotic death process (see also Chapter 23). The unique structure of mitochondria is important in allowing them to perform these necessary cellular functions.

A. Function in energy production

One characteristic feature of mitochondria is the double phospholipid bilayer membrane that forms their outer boundary (Figure 5.6). The inner mitochondrial membrane forms folded structures called **cristae** that protrude into the mitochondrial lumen (space) known as the mitochondrial **matrix**. Protons (H⁺) are pumped out of the mitochondrial matrix, creating an electrochemical gradient of protons. The flow of protons back into the matrix drives the formation of ATP from carbohydrates and lipids in the process of **oxidative phosphorylation** (see also *LIR Biochemistry*, pp. 77–80). The presence of mitochondria within a cell enhances the amount of ATP produced from each glucose molecule that is broken down, as evidenced by human red blood cells that lack mitochondria. In red blood cells, only 2 ATP molecules are generated per glucose molecule. In contrast, in human cells with mitochondria, the yield of ATP is as high as 32 per glucose molecule.

B. Role as independent units within the eukaryotic cells

Mitochondria also contain DNA (mtDNA) and ribosomes for the production of RNA and some mitochondrial proteins. mtDNA is approximately 1% of total cellular DNA and exists in a circular arrangement within the mitochondrial matrix. Mutations or errors in some mitochondrial genes can result in disease. Most mitochondrial proteins, however, are encoded by the genomic DNA of the cell's nucleus. Mitochondria divide by fission, as do bacteria, and are actually believed to have arisen from bacteria that were engulfed by ancestral eukaryotic cells.

C. Function in cell survival

Survival of eukaryotic cells depends on intact mitochondria. At times, the death of an individual cell is important for the benefit of the organism. During development, some cells must die to allow for proper tissue and organ formation. Death of abnormal cells, such as virally infected cells or cancerous cells, is also for the good of the organism. In all these cases, mitochondrial involvement is important to ensure cell survival when appropriate and also to facilitate programmed cell death when necessary. When the process of programmed cell death or apoptosis is stimulated in a cell, proapoptotic proteins insert into the mitochondrial membrane, forming pores. A protein known as **cytochrome c** can then leave the intermembrane space of the mitochondria through the pores, entering the cytosol (Figure 5.7). Cytochrome c in the cytosol stimulates a cascade of

Clinical Application 5.1: Mitochondrial Diseases

Mitochondrial cytopathies are disorders that result in an inability of mitochondria to properly produce ATP. These disorders may result from mutations in mtDNA or from mutations in genomic genes that encode mitochondrial proteins and enzymes. Because mitochondria from sperm cells do not enter a fertilized egg, mitochondria are inherited exclusively from the mother. Therefore, disorders of mtDNA are also inherited from the mother only. Siblings share mitochondria with each other and with their mother, causing mitochondrial disorders that arise from mtDNA mutations to occur within families. Some individuals may be affected more or less severely even within a family. It is estimated that 1 in 4,000 children in the United States will develop a mitochondrial disorder by age 10. Some diseases of aging (type 2 diabetes, Parkinson disease, Alzheimer disease, atherosclerosis, etc.) may also result in part from decreased mitochondrial function.

More than 40 different mitochondrial disorders are described. They share the common feature of a reduced ability of mitochondria to completely oxidize or breakdown fuel sources such as carbohydrates. The buildup of intermediates can further damage mitochondria and mtDNA, which does not have an efficient repair mechanism. Mitochondrial diseases are categorized by the organ that is affected. Defects in oxidative phosphorylation will affect tissues with the greatest need for ATP. Brain, heart, liver, skeletal muscles, and eyes are examples of organs often affected in some mitochondrial cytopathies. Developmental delays, poor growth, loss of muscle coordination, and loss of vision are various signs of these disorders.

Kearns-Sayre syndrome is an example of a mitochondrial disorder caused by defective mtDNA. It is rare and results in paralysis of eye muscles and degeneration of the retina. A single large deletion of mtDNA is responsible for the development of this syndrome. Leber hereditary optic neuropathy results in blindness, primarily in young men. A single change (point mutation) in mtDNA causes this disorder. Deletions in mtDNA can result in Pearson syndrome, where there are bone marrow and pancreas dysfunctions. Cures are not presently available for mitochondrial cytopathies and treatments are designed to reduce symptoms or to prevent progression of disease.

biochemical events resulting in apoptotic death of the cell (see also Chapter 23).

IV. LYSOSOMES

Lysosomes are membrane-enclosed organelles of various sizes that have an acidic internal pH (pH 5) (Figure 5.8). They are formed from regions of the Golgi complex that pinch off when proteins destined for the lysosome reach the trans Golgi (see also Chapter 11). Lysosomes contain potent enzymes known collectively as **acid hydrolases**. These enzymes are synthesized on ribosomes bound to the ER. They function within the acidic environment of lysosomes to hydrolyze or break down macromolecules (proteins, nucleic acids, carbohydrates, and lipids). Lysosomes play a critical role in the normal turnover of macromolecules

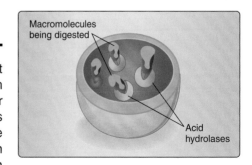

Figure 5.8
Lysosome structure and function.

that have reached the end of their functional life. Nonfunctional macro-molecules build up to toxic levels if they are not degraded within lyso-somes and properly recycled for reuse within the cell. This is exemplified by diseases known as lysosomal storage diseases. Such diseases are caused by defective acid hydrolases, resulting in accumulation of sub-strates of the defective acid hydrolases. Most are fatal at an early age. In infantile Tay-Sachs disease, gangliosides accumulate in the brain and death occurs by age 4. In addition to degrading cellular macromol-ecules at the end of their life span, lysosomal enzymes also degrade materials that have been taken up by the cell through endocytosis or phagocytosis.

Clinical Application 5.2: Lysosomal Storage Diseases

Lysosomal storage diseases are caused by defects in acid hydrolases. (An exception is I-cell disease, in which acid hydrolases do not traffic properly to the lysosomes.) Over 40 different acid hydrolases exist within normal, healthy lysosomes. The absence of particular acid hydrolases can lead to the accumulation of particular macromolecule substrates within the lyso-somes. Therefore, these substances build up within the lysosomes instead of being degraded and recycled. Lysosomal storage diseases are catego-rized by the type of compound that builds up to toxic levels within the lyso-somes. For example, gangliosides accumulate in Tay-Sachs disease, and glycosaminoglycans (also known as mucopolysaccharides) accumulate in "mucopolysaccharidoses" such as Hurler syndrome and Hunter syndrome. Both Hunter and Hurler syndromes can be severe, with hearing loss and damage to the central nervous system. Children with Hurler syndrome usu-ally stop developing between 2 and 4 years of age. Individuals with early-onset Hunter syndrome usually have a life span of 10 to 20 years. In Farber disease, ceramide accumulates as a result of acid ceramidase deficiency and is fatal within the first year of life. Some other lysosomal storage dis-eases do not become evident until much later in life. The adult-onset form of Gaucher syndrome (type I) is the most common lysosomal storage dis-ease. This disease results from a deficiency of glucosylceramidase and results in glucosylceramide lipidosis (excess of this particular type of lipid). Splenomegaly (enlargement of the spleen) and bone pain are characteris-tics. The infantile form of Gaucher syndrome (type II) is much more severe, with neurological impairment and death by age 3.

Clinical Application 5.3: Tay-Sachs Disease

Three forms of Tay-Sachs disease are recognized: infantile, juvenile, and adult/late onset. The disease is characterized by the accumulation of gan-gliosides in the brain owing to low activity or complete deficiency of the lysosomal acid hydrolase β-hexosaminidase A. Accumulation occurs sooner or later in life depending on the extent of enzyme activity retained by the affected person. Individuals with the infantile form usually die from the disease between ages 2 and 4, while those with the juvenile form live to between 5 and 15 years, with progressive motor skill deterioration. Those with adult-/late-onset form have speech and swallowing difficulties, cognitive decline and progressive neurological deterioration, psychiatric illness, and gait disturbances but often do not die as a result of the disorder.

The infantile form of Tay-Sachs disease is the most common variant and is inherited as an autosomal recessive disorder. It results when severe mutations in the *HEXA* gene on human chromosome 15 are inherited from both parents, resulting in absence of β-hexosaminidase A enzyme activity. (In other forms of Tay-Sachs disease, mutations may result in decreased activity, but not total absence of β-hexosaminidase A.) More than 100 different mutations are known in this gene and different mutations are seen in different populations. Individuals may also inherit different mutations from each parent (compound heterozygote). Fewer than 30 children are born each year in the United States with infantile Tay-Sachs disease. Children with this disorder are normal at birth but develop signs of the disease by about 6 months of age. Affected individuals have a particularly strong response to sudden noises ("startle response") and may have hypertonia. Signs and symptoms occur in response to neurons becoming distended with gangliosides that have accumulated abnormally. Deterioration of mental and physical abilities happens quickly and may include inability to swallow, blindness, deafness, and paralysis, with death often by age 2 and generally before age 4.

Overall, the infantile form of the disease is seen in approximately 1/320,000 births in the United States, but to 1/3,600 births to couples of Ashkenazi Jewish background. These figures correlate with approximately 1/250 healthy persons in the general United States population being carriers and to between 1/27 and 1/30 healthy persons of Ashkenazi descent being carriers. Cajuns in southern Louisiana have similarly high carrier risk to the Ashkenazi population. French Canadians in southeastern Quebec have a similarly high carrier risk of approximately 1/27 but carry different mutations in *HEXA* from those commonly seen in Ashkenazi and Cajun populations. New estimates are that as many as 1/50 Americans of Irish descent may also be carriers of *HEXA* mutations.

V. PEROXISOMES

Peroxisomes resemble lysosomes in size and in structure. They have single membranes enclosing them and contain hydrolytic enzymes. However, they are formed from regions of the ER as opposed to regions of the Golgi complex. Enzymes that function in peroxisomes are synthesized on free ribosomes and are not modified in the ER or Golgi complex. Within peroxisomes, fatty acids and purines (AMP and GMP) are broken down (see also Chapter 7 and *LIR Biochemistry*, p. 195). Hydrogen peroxide, a toxic by-product of many metabolic reactions, is detoxified in peroxisomes. Within liver cells (hepatocytes), peroxisomes participate in cholesterol and bile acid synthesis (see also *LIR Biochemistry*, pp. 220–224). Peroxisomes are also involved in the synthesis of **myelin**, the substance that forms a protective sheath around many neurons.

Some rare inherited diseases are caused by impaired peroxisome function. They exert their effects from birth onward and life expectancy is short. For example, X-linked adrenoleukodystrophy (the disease of the young boy in the 1992 film, *Lorenzo's Oil*) is characterized by the deterioration of myelin sheaths of neurons, owing to the failure of proper fatty acid metabolism. Zellweger syndrome is caused by a defect in the transporting of peroxisomal enzymes into the peroxisomes in liver, kidneys, and brain. Affected individuals do not usually survive beyond 6 months of age.

Chapter Summary

- Organelles are intracellular structures within eukaryotic cells that are responsible for carrying out specific functions necessary for normal cellular life.
- The nucleus, ribosomes, and endoplasmic reticulum (ER) function together in processing proteins that will function outside of the cell or within lysosomes.
- The nucleus is enclosed by a double-membrane layer and houses the genomic DNA of the cell within the chromosomes.
- The nucleolus within the nucleus is the site of ribosome manufacture.
- Ribosomes function in protein translation and may be free or else bound to the ER.
- The ER is contiguous with the nuclear envelope and a series of membrane-enclosed spaces in which protein processing can occur.
- Rough endoplasmic reticulum has attached ribosomes while smooth endoplasmic reticulum does not.
- The Golgi complex appears like a series of flat, membrane-enclosed sacs with three distinct regions (cis, medial, and trans). It is involved in modifying and packaging of new proteins.
- Mitochondria have double membranes that form folded cristae and surround a matrix.
- ATP is generated using an electrochemical gradient that exists across the matrix.
- Mitochondria can self-replicate and contain their own DNA and ribosomes. They are believed to have arisen from bacteria engulfed by ancestral eukaryotic cells.
- Cell survival depends on the integrity of the mitochondrial membrane. When pores are placed in the membrane, cytochrome c is released into the cytosol, setting off a cascade of reactions that lead to programmed cell death.
- Lysosomes contain powerful digestive enzymes known as acid hydrolases that function within their acidic environment.
- Defects in lysosomal acid hydrolases cause lysosomal storage diseases where accumulation of nonfunctional macromolecules causes cellular damage and death at an early age.
- Peroxisomes contain hydrolytic enzymes, detoxify hydrogen peroxide, and participate in the breakdown of fatty acids. They are involved in liver synthesis of cholesterol and in the production of myelin sheaths that protect neurons.

Study Questions

Choose the ONE best answer.

5.1 A cytosolic cellular structure with two subunits is observed to assemble and disassemble and to bind to mRNA and to associate, at times, with endoplasmic reticulum. The most likely identity of this structure is a/an

A. Golgi complex.
B. Lysosome.
C. Nucleus.
D. Peroxisome.
E. Ribosome.

Correct answer = E. Ribosomes are composed of two subunits and exist within the cytosol, often bound to endoplasmic reticulum (ER). They participate in protein translation from mRNA and bind to mRNA during the process. Golgi, lysosomes, nuclei, and peroxisomes are not formed of subunits that assemble and disassemble. None bind to mRNA or to ER.

5.2 A single membrane-enclosed organelle is observed to be in close proximity to the plasma membrane. It appears to surround newly modified proteins in membrane-enclosed structures. The most likely identity of this organelle is

A. Golgi complex.
B. Lysosome.
C. Mitochondria.
D. Nucleus.
E. Peroxisome.

Correct answer = A. The Golgi complex is a series of membrane-enclosed tubules involved in protein processing. It is localized near the plasma membrane and places newly modified proteins within the vesicles that bud off from the Golgi. Mitochondria and nuclei have double membranes. Lysosomes and peroxisomes do have single membranes but are not involved in protein processing. Lysosomes are produced from the Golgi and peroxisomes from the endoplasmic reticulum.

5.3 A membrane-enclosed intracellular structure is observed to release a protein through a pore into the cytosol. Following this release, biochemical reactions take place and result in the cell's death by apoptosis. The most likely identity of this intracellular structure is

A. Golgi complex.
B. Lysosome.
C. Mitochondria.
D. Nucleus.
E. Peroxisome.

Correct answer = C. Mitochondria release cytochrome *c* into the cytosol, initiating a cascade of biochemical events that result in apoptotic cell death. The nucleus contains the cell's DNA. The Golgi complex participates in modifying and sorting newly produced proteins. Lysosomes and peroxisomes are distinct from each other but are both involved in digestion.

5.4 An organelle with DNA, distinct from genomic, chromosomal DNA and ribosomes is believed to have originally been a single-celled independent organism engulfed by ancestral eukaryotic cells. This organelle is

A. A mitochondrion.
B. The cis Golgi complex.
C. Rough endoplasmic reticulum.
D. A nucleus.
E. A peroxisome.

Correct answer = A. A mitochondrion is an organelle with its own DNA that is separate from the genomic DNA of the nucleus. Mitochondria are believed to have been engulfed by ancestral eukaryotic cells. The cis Golgi is a portion of flattened sacs between the rough ER and medial Golgi and contains no DNA. Rough ER have attached ribosomes and function in translation of protein from mRNA. Nuclei contain genomic DNA within chromosomes. Peroxisomes are bounded by single membranes and breakdown fatty acids and purines and detoxify hydrogen peroxide.

5.5 Nuclear lamina is composed mainly of

A. Collagen.
B. Microtubules.
C. Phospholipids.
D. Cholesterol.
E. Intermediate filaments.

Correct answer = E. Nuclear lamina is composed mainly of intermediate filaments. The interior of the nucleus containing the nucleoplasm is organized by the nuclear lamina, the protein scaffolding of the nucleoplasm that is composed mainly of intermediate filaments, which are cytoskeletal components. Microtubules are also cytoskeletal components but not the main ones found in nuclear lamina. Collagen is secreted by cells into the extracellular matrix and is not found within the nucleus. Phospholipids and cholesterol are main constituents of biological membranes.

5.6 An organelle is bounded by a single membrane and contains hydrolytic enzymes that were synthesized on free ribosomes. From which structure was this organelle derived?

A. Mitochondria
B. Nucleus
C. Endoplasmic reticulum
D. Golgi
E. Lysosome

Correct answer = C. The organelle described is a peroxisome, which is derived from regions of the endoplasmic reticulum, and not from regions of the Golgi complex, which form lysosomes. Mitochondria and nuclei do not directly give rise to the structures of other organelles.

5.7 Two-cell types are compared for their ability to produce ATP from glucose, and while one produces only 2 ATP per glucose, the other has an ATP yield of 32 ATP from the glucose. The most likely difference between these cells that contributes to this finding is

A. Mutation in acid hydrolase within lysosomes of one cell type only.
B. Presence of mitochondria in one cell type but not in the other.
C. Absence of a particular peroxisomal hydrolase in one cell type only.
D. One cell types' inability to breakdown fatty acids and purines to extract energy.
E. Accumulation of glycosaminoglycans in one cell type, inhibiting ATP production.

Correct answer = B. The presence of mitochondria within a cell enhances the amount of ATP produced per glucose. One cell type was likely to be red blood cells, which lack mitochondria and produce only 2 ATP per glucose compared with 32 ATP per glucose in cells with mitochondria. This mitochondrial-dependent process of ATP production in mitochondria is oxidative phosphorylation. It is not dependent on peroxisomes or lysosomes (or their acid hydrolases) or breakdown of fatty acids and purines, a process that normally occurs in peroxisomes. Glycosaminoglycans accumulate in particular lysosomal storage diseases, unrelated to ATP production from mitochondria.

5.8 A 7-month-old girl who previously had a normal development now exhibits signs and symptoms of Tay-Sachs disease. Which of the following organelles is affected in this disorder?

    A. Endoplasmic reticulum
    B. Golgi
    C. Lysosomes
    D. Mitochondria
    E. Peroxisomes

> Correct answer = C. Tay-Sachs disease is a lysosomal storage disease. A mutant lysosomal hydrolase prevents the breakdown of certain macromolecules. In Tay-Sachs disease, gangliosides accumulate in the brain, causing the signs and symptoms. The other organelles listed are unaffected in Tay-Sachs disease.

5.9 The child with Tay-Sachs disease, described in question 5.8, most likely acquired this disorder via

    A. Inheritance of mutation on X chromosome inherited from her father.
    B. Autosomal dominant inheritance of mutant *HEXA* gene from one of her parents.
    C. Mitochondrial inheritance of *HEXA* mutation from her mother.
    D. Inheritance of the same or different *HEXA* mutations from both her parents.
    E. Acquisition of defective peroxisomes via engulfment.

> Correct answer = D. Tay-Sachs disease in an infant is acquired via autosomal recessive inheritance, with one mutant *HEXA* gene from each parent. The *HEXA* gene is on chromosome 15, not on the X chromosome or encoded by mitochondrial DNA. Inheritance is autosomal recessive, not autosomal dominant. The disorder is lysosomal, not peroxisomal; peroxisomes are not engulfed by cells but are intracellular organelles.

5.10 A previously healthy 24-year-old male develops optic neuropathy and becomes blind. His mother's brother has the same condition. The patient is told that while his condition is inherited, there is no chance that he will pass his mutant gene to any of his future children. What type of disorder most likely affects this patient?

    A. Autosomal recessive disorder
    B. Disease of aging
    C. Lysosomal storage disease
    D. Mitochondrial disease
    E. Peroxisomal disorder

> Correct answer = D. This patient most likely has a mitochondrial disease. (He may have Leber hereditary optic neuropathy.) Mitochondrial diseases are inherited exclusively from the mother. Since sperms do not enter fertilized eggs, a male individual cannot pass on a defective mitochondrial gene to his children. This is the only type of condition that a male has no chance of passing on to any of his children. Autosomal recessive genes require both parents to pass on a mutant gene in order for a child to be affected. In autosomal conditions though, an affected person would have a 50% chance of passing on a mutant gene to his children. A 24-year-old man is not old enough to exhibit signs and symptoms of a disease of aging. Lysosomal storage diseases are often autosomal recessive. Peroxisomal disorders are generally present at birth and affected individuals have a very short life expectancy.

UNIT II

Organization of the Eukaryotic Genome and Gene Expression

They are in you and me; they created us, body and mind; and their preservation is the ultimate rationale for our existence… they go by the name of genes, and we are their survival machines.
—Richard Dawkins (English biologist, 1941–)
In: *The Selfish Gene* (1976)

Human genetic information, collectively known as the genome, exists as deoxyribonucleic acid or DNA within every nucleated somatic cell of the body. The DNA in each cell contains all instructions necessary to direct the growth and development of cells into an organism and to maintain cellular function throughout the individual's lifespan. Replication or copying of DNA, during development and also during growth and repair, must ensure that the instructions within DNA are faithfully passed from cellular generation to generation. This requirement ensures the preservation of both the individual organism and the species. Neither human DNA nor human cells can exist without the other, in a symbiotic type of relationship. Cells provide the framework and machinery to ensure that the genetic instructions within the DNA are reproduced and followed with fidelity.

Within DNA are nucleotide bases that are arranged in specific sequences to form genes. Genes exist to code for proteins that carry out functions of the cells and, therefore, of the organism. Yet, while this basic genetic blueprint is identical within all somatic cells of an individual, proteins within cells differ according to the cell type. While liver cells, for example, require an array of protein enzymes to carry out metabolic functions, bone cells require more protein support structures. Through transcription, instructions of DNA are converted into mRNA and the nucleic acid language is then translated into a protein. By regulating the regions of DNA that are transcribed, specific cell types are able to dictate the proteins that are produced. After their manufacture, the proteins are trafficked or moved to functional locations within or outside their cell or origin. A careful balance is needed between protein synthesis and degradation to enable cell function and survival and, therefore, also the continued existence of the DNA that instructs its formation.

6 The Eukaryotic Genome

I. OVERVIEW

Every nucleated eukaryotic somatic cell contains essentially the same blueprint—a set of genetic information collectively known as the **genome**. The tremendous potential for understanding the genetic basis for human development and disease led to the successful worldwide effort to sequence the entire human genome. We still need to understand how the human genome is organized and how to decipher the meaning of much of the human deoxyribonucleic acid (DNA) sequence.

Some parts of the genome contain instructions used daily by the cell. But other genetic instructions are useful to a cell only when it is stressed. Still other genetic instructions are never used by the cell. Because of the vast amount of genetic information, it is critical for a cell to retrieve this information in a timely manner. Knowledge of how genetic information is stored and retrieved is essential for an understanding of the functioning of the eukaryotic genome. We will first examine the physical organization of the genome and then proceed to the biochemical processes required to maintain and manage the genome. Just as the page you are reading contains letters that are arranged in discrete information units known as words and words are combined into sentences, paragraphs, chapters, and so on, DNA contains nucleotides arranged into genes, chromosomes, and so on (Figure 6.1). This chapter will describe both the physical and the informational organization of the genome.

II. PHYSICAL ORGANIZATION

The human genome is contained within two distinct compartments: the nucleus and the mitochondria. The bulk of the genome, containing about 20,000 to 25,000 genes encoded by DNA, is contained within a set of linear chromosomes within the cell nucleus and contains genetic material of both maternal and paternal origin. In contrast, mitochondrial DNA contains 37 genes that are essential for normal mitochondrial function and is exclusively of maternal origin. This chapter addresses the organization of nuclear DNA. Eukaryotic nuclear DNA is associated with a variety of proteins, which together form a complex structure, **chromatin**, that allows for numerous configurations of the DNA molecule and types of control unique to the eukaryotic organism.

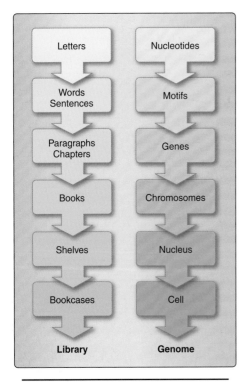

Figure 6.1
Data storage analogy.

A. DNA building blocks

DNA contains the structural blueprint for all genetic instructions. The genetic code contained within the DNA is composed of four "letters" or bases. Two of the bases are heterocyclic compounds or purines—adenine (A) and guanine (G)—and two are six-member rings known as pyrimidines—cytosine (C) and thymidine (T). The famous double-helix structure of DNA derives from its phosphate-deoxyribose backbone (Figure 6.2). The backbone comprises five-carbon sugar (pentose) molecules bound to a nucleoside (A, G, C, or T). The pentose molecules are also asymmetrically joined to phosphate groups by phosphodiester bonds. Hydrogen bonds between complementary (G:C or A:T) nucleotides (a nucleoside linked to a sugar and one or more phosphate groups) interact to stabilize and form the double-helix structure.

B. Histones

Chromatin consists of very long double-stranded DNA molecules, nearly an equal mass of rather small basic proteins termed **histones**, as well as smaller amounts of nonhistone proteins, and a small quantity of ribonucleic acid (RNA). Histones are a heterogeneous group of closely related arginine- and lysine-rich basic proteins, which together make up one-fourth of amino acid residues. These positively charged amino acids help histones to bind tightly to the negatively charged sugar phosphate backbone of DNA. Functionally, histones provide for the compaction of chromatin. However, chromatin is far from static and can be altered in a dynamic fashion with changes to differentiation status of the cell.

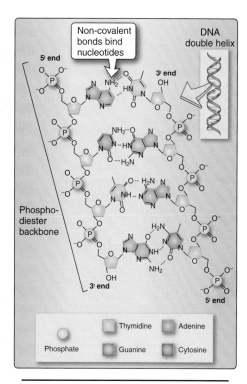

Figure 6.2
Nuclear structure of eukaryotic DNA.

> ### Just how much DNA is there?
>
> The human haploid genome contains approximately 3×10^8 base pairs packaged into 23 chromosomes. Total uncoiled DNA within a single human cell would stretch to more than a meter. Uncoiled individual chromosomes would measure 1.7 to 8.5 cm in length.

C. DNA packaging

The nucleus of a human cell is typically 6 μm in diameter but contains DNA that at its maximum stages of condensation is only about 1/50,000th of its linear length. At least four levels of packaging of DNA take place in order that DNA in individual chromosomes fits into the 1.4-μm chromosome seen at metaphase (a stage in mitosis in the cell cycle where the DNA is most condensed).

Nucleosomes are the fundamental organization upon which the higher-order packing of chromatin is built. Each nucleosome core consists of a complex of eight histone proteins (two molecules each of histone H2A, H2B, H3, and H4) with double-stranded DNA wound around it. 146 base pairs (bp) of DNA are associated with the nucleosome particle, and a 50- to 70-bp span of linker DNA bound by a linker histone H1 separates each nucleosome (Figure 6.3).

In addition to their role in packaging DNA, nucleosomes also regulate gene expression, or activity, by determining whether the DNA sequences can be accessed by transcription factors, allowing the factors to regulate expression of a nearby gene (Chapter 10).

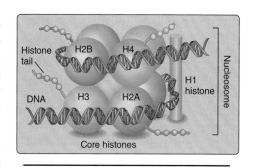

Figure 6.3
Structure of a nucleosome.

Nucleosomes are in turn successively packed into higher-order structures by coiling and looping (Figure 6.4).

D. Histone modification

Each core histone has a structured domain and an unstructured amino-terminal "tail" of 25 to 40 amino acid residues (see figure 6.3). Enzymatic modification of the amino-terminal tails (e.g., by acetylation,

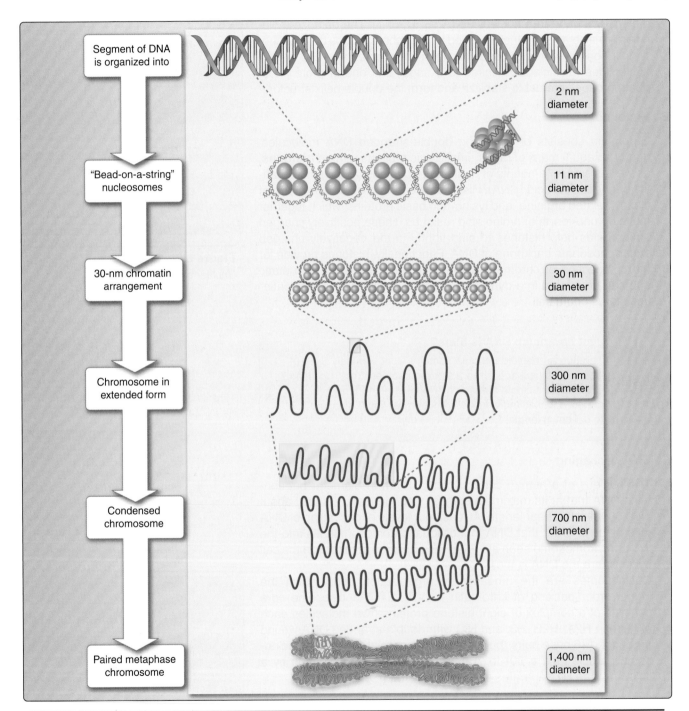

Figure 6.4
Higher-order structures formed during progressive compaction of chromatin.

methylation, or phosphorylation) modifies the histones' net electric charge and shape. These modifications are physiologically reversible and are thought to prepare the chromatin for DNA replication and transcription. (See more under "Epigenetics.")

1. **Euchromatin and heterochromatin:** These terms describe the compaction of DNA in the chromosome and are used to further classify chromatin. Densely packed or compacted regions of chromatin are termed **heterochromatin** and, for the most part, are genetically inactive (Figure 6.5). Transcription is inhibited in heterochromatin because the DNA is packaged so tightly that it is inaccessible to the proteins responsible for RNA transcription.

2. **Transcriptionally active nucleus:** Less densely compacted chromatin regions in a transcriptionally active nucleus are called **euchromatin** (Figure 6.5) and are commonly undergoing, or preparing for, or have just completed transcription. For a gene to be transcribed, its gene sequence must become available to the RNA polymerases and regulatory proteins that influence the rate at which the gene is transcribed. Euchromatin represents uncoiled chromatin structures that allow RNA polymerases' and regulatory proteins' access to DNA. During cell division, the chromatin becomes highly compact and coiled and condenses into the familiar structure of the mitotic chromosome.

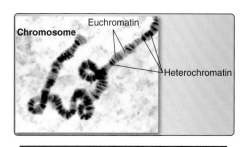

Figure 6.5
Euchromatin and heterochromatin.

E. Chromosome structure

Individual chromosomes are composed of both a noncovalent complex of one very long, linear duplex DNA and associated histone proteins. Chromosome structure varies with the cell cycle, from the loose thread-like appearance in G_1 phase to the tightly compacted state observed during M phase (see Chapter 20). Chromosomes require three sequence elements for their propagation and maintenance as individual units. **Telomeres** are hexameric DNA repeats [$(TTAGGG)_n$] found at the ends of chromosomes that serve to protect the chromosome from degradation (Figure 6.6). Sequence elements known as **centromeres** serve as "handles," which allow mitotic spindles to attach to the chromosome during cell division. As the cell progresses through the mitotic or M phase of the cell cycle, the nuclear envelope breaks down, and chromosomes segregate into the opposite poles of the cells (to form daughter cells), while a **kinetochore** forms consisting of the centromere and mitotic spindles. The centromere also serves as a boundary that separates the two arms (short, or **p**, from the French *petite*, and long, or **q**, because "q" follows "p" in the alphabet) of the chromosome (placement varies for different chromosome types). We discuss more about the mechanism of the cell cycle in Chapter 20.

In order for DNA in chromosomes to replicate, a specific nucleotide sequence acts as a DNA replication origin. Each chromosome contains **multiple origins of replication**, dispersed throughout its length. At the origin of replication, there is an association of sequence-specific, double-stranded DNA-binding proteins with a series of direct repeat DNA sequences.

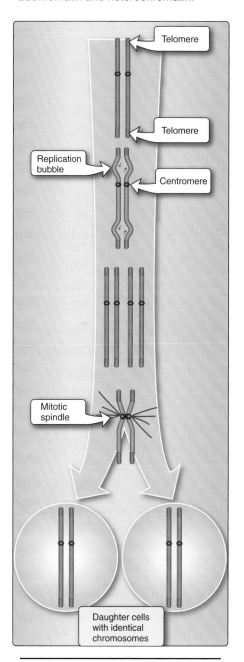

Figure 6.6
Chromosome structure.

Clinical Application 6.1: Karyotype Analysis

Metaphase chromosomes can be visualized microscopically to allow geneticists to detect and identify chromosome abnormalities. Karyotype analysis is an important diagnostic tool in the prenatal diagnosis of chromosomal abnormalities such as Down syndrome (trisomy 21), in the staging of tumor progression (tumors often have abnormal number of chromosomes), in determining infertility, and even to prevent males from performing as athletes in female sports events.

III. INFORMATION ORGANIZATION

Collectively, **ploidy** refers to the number of chromosome copies within a cell. Most of the somatic cells within the body are **diploid**, meaning that each nucleus has two copies of each chromosome, one deriving from the mother and other from the father. Germ cells are the exception to this rule, which contain a single copy of each chromosome and are known as haploid.

The haploid genome of each human cell consists of 3.0×10^9 bp of DNA, divided into 23 (22 somatic and 1 sex) chromosomes. The entire haploid genome contains sufficient DNA to code for nearly 1.5 million pairs of genes. Surprisingly, the human genome project has shown humans to have only about 20,000 to 25,000 genes. Our genome has approximately the same number of genes as a fruit fly, yet considerably more complex than that organism. Our protein-coding genes appear to produce more than one protein product by alternative splicing (Chapter 10). The human **proteome**, or total number of protein species, is five to ten times larger than that of the fruit fly.

A new class of genes, **microRNAs**, has recently been discovered and appears to regulate at least 30% of all proteins within the human proteome (see Chapters 8 and 10). MicroRNAs are involved in many human diseases, ranging from diabetes, obesity, and viral diseases to various types of cancer.

Genomic, eukaryotic DNA may be further classified as either unique or single copy or as a repetitive sequence DNA (Figure 6.7).

A. Unique DNA sequences

Single-copy DNA or genes generally encode information for specific protein products. The 20,000 to 25,000 genes within the human genome can be divided into four general categories. Approximately 5,000 genes are involved with genome maintenance, nearly 5,000 with signal transduction, and 4,000 with general biochemical functions; the largest portion, 9,000 genes, is involved in other activities (Figure 6.8).

B. Repeat sequences

Although repeat sequences do not encode proteins, they make up at least 50% of the human genome. These sequences do not appear to have direct functions but are important for chromosome structure and

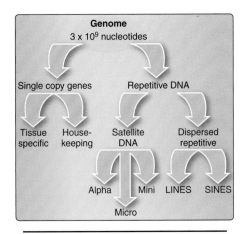

Figure 6.7
Organization of the genome.

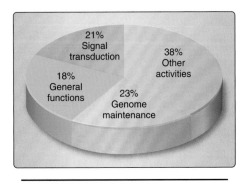

Figure 6.8
Unique (nonrepetitive) DNA distribution with the genome.

dynamics. These sequences fall into two main classes: Satellite DNA and LINES and SINES.

1. **Satellite DNA:** These highly repetitive sequences tend to be clustered and repeated many times in tandem (a head-to-toe arrangement). They are generally not transcribed and are present in 1 to 10 million copies per haploid genome. These sequences are also associated with the centromeres and telomeres of chromosomes.

   Satellite DNA sequences are categorized according to the number of base pairs within the repeat sequence:

   - Alpha satellite—171 bp sequence that extends several million base pairs or more in length
   - Minisatellite—20 to 70 bp in length and a total length of a few thousand base pairs
   - Microsatellite—repeat units only 2, 3, or 4 bp in length and a total length of a few hundred

---

Clinical Application 6.2: Microsatellites and Minisatellites and Genetic Mapping

Repeat DNA sequences are widespread throughout the human genome and are polymorphic (a common genetic variation in nucleotide sequence). Consequently, they have found application as genetic markers for identity testing and disease diagnosis. Short tandem repeats (STRs) are microsatellites of 2 to 6 bp in length and are important for forensic laboratories because these sequences can be readily amplified with the polymerase chain reaction (PCR) from small amounts of suboptimal-quality DNA. Formerly, minisatellite sequences were identified by restriction fragment length polymorphism (RFLP) analyses, a more cumbersome technique requiring larger amounts of moderately good quality DNA to successfully amplify DNA.

In the United States, a core set of 13 STR markers is currently used to generate a nationwide DNA database called the FBI Combined DNA Index System (CODIS). The CODIS and similar DNA databases have successfully linked DNA profiles from repeat offenders and crime scene evidence. STR typing has also found application in the resolution of paternity disputes.

Trinucleotide repeats are microsatellite sequences that are normally present in certain genes and can undergo expansion. Several human diseases have been shown to result from the expansion of the number of repeats above the normal number resulting in an unstable and defective gene (Table 6.1).

2. **LINES and SINES:** These unclustered sequences are found interspersed with unique sequences. These are also present at less than 10^6 copies per haploid genome. They are transcribed into RNA and can be grouped according to their size.

   - LINES (**l**ong **in**terspersed **e**lement**s**) 7,000 bp (20 to 50,000 copies)
   - SINES (**s**hort **in**terspersed **e**lement**s**) 90 to 500 bp (about 100,000 copies)

Table 6.1: Trinucleotide Repeats and Disease

Disease	Repeat Sequence	Normal Number	Disease-State Number	Location
Kennedy	CAG	11–33	40–62	Protein-coding region
Huntington	CAG	11–34	42–100	Protein-coding region
Fragile X	CGG	6–54	250–4,000	5′ untranslated region
Myotonic dystrophy	CTG	5–30	>50	5′ untranslated region

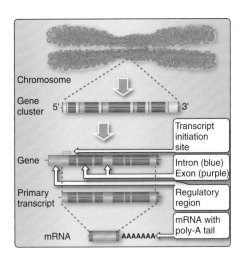

Figure 6.9
Gene organization.

IV. FUNCTIONAL ORGANIZATION

Functions within the cell are usually encoded by genes. Genes are present on the chromosomes and in the mitochondria. Not all genes are active in all tissues and specific alterations to these genes are necessary for tissue-specific gene expression.

A. Genes

A **gene** is the complete sequence region necessary for generating a functional product. This encompasses promoters and control regions necessary for the transcription, processing, and, if applicable, translation of a gene. About 2% of the genome encodes instructions for the synthesis of proteins. Genes appear to be concentrated in random areas along the genome, with vast expanses of noncoding DNA between them (Figure 6.9).

B. Epigenetics

Aside from mutation, all cells in an individual have identical DNA content and sequence. However, different tissues and cells require a specific set of genes to carry out their functions. Structural modification that differs among cell types plays an important role in the control of gene expression during development and differentiation. These changes do not affect the DNA sequence of the genome and are termed **epigenetic**. These changes commonly occur in concert and help explain alterations in gene expression that are stably inherited. The following represent epigenetic mechanisms operative in eukaryotic cells.

1. **Methylation of cytosines:** Specific cytosine methylation correlates with gene expression and is achieved through processes of adding or removing methyl groups from nucleotides in DNA. DNA methylation is involved in a wide variety of fundamental cellular processes. The major site of DNA methylation in mammals is on a cytosine base in DNA—especially the 5′ cytosine adjacent to a guanosine base (5′-CG-3′) (Figure 6.10).

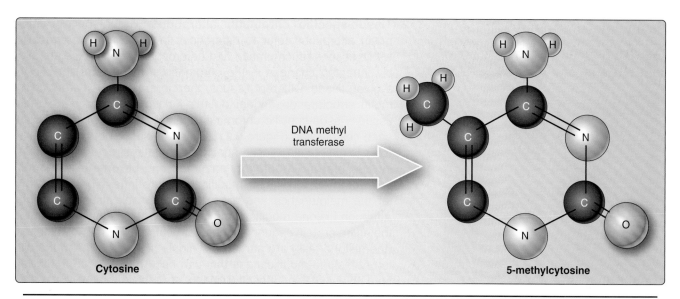

Figure 6.10
Methylation of specific cytosines in DNA.

Gene methylation

5′-CG-3′ residues tend to cluster in the promoter regions of genes. Sometimes they are constitutively hypomethylated and active in all cell types and referred to as housekeeping genes. Tissue-specific genes tend to be preferentially methylated in cells/tissues where they are not needed. One example is the globin gene. The globin gene is actively synthesized in hematopoietic cells (cells that give rise to the different blood types) that give rise to reticulocytes (newly formed red blood cells) while the same gene is methylated and silenced in tissues that are not required to synthesize globin protein. Methylation of 5′-CG-3′ residues in DNA is thought to cause steric hindrance to the binding of proteins that influences gene expression.

Clinical Application 6.3: Genomic Imprinting

Although most genes have equal representation from both parents, some genes appear to be expressed exclusively from the chromosome contributed from either the mother or the father. The silencing of genes on these chromosomes is a result of gene methylation. These genes are said to be "imprinted" or have the ability to be turned on or off depending on which parent contributed the gene. The deletion of a certain region of chromosome 15 produces different outcomes depending on which parent contributed it. A deletion of this region derived from the father causes Prader-Willi syndrome, while deletion of the same region from the mother's chromosome produces Angelman syndrome. These conditions have very little in common in terms of pathology, even though the same area of the chromosome undergoes deletion in both. In the case of Prader-Willi syndrome, the failure to express several genes from the paternal chromosome causes the disease, while it is the failure of the maternal chromosome to express a different gene that results in Angelman syndrome.

All cells within an organism have the same DNA but different chromatin structure, which determines tissue-specific function. Kidney and liver are shown as examples below.

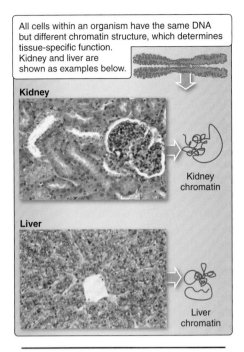

Kidney

Kidney chromatin

Liver

Liver chromatin

Figure 6.11
Chromatin structure is tissue specific.

2. **Tissue-specific chromatin alterations:** Tissue-specific chromatin alterations refer to differences in the chromatin structure within tissues (euchromatin and heterochromatin). These are stably inherited changes in the chromatin structure and specific to a given tissue. Different tissues, therefore, have a different chromatin structure dependent on the functions carried out by the tissue. Tissue-specific chromatin alterations are maintained with each cell division (Figure 6.11). Different classes of proteins play roles in maintaining tissue-specific chromatin structure. Examples include **histone-modifying enzymes** and **chromatin remodeling complexes**. Histone modifications are inheritable and allow for the maintenance and propagation of stable tissue structure. Although histone modifications can occur at locations throughout the entire protein sequence, the unstructured N-termini of histones (the histone tails) are particularly highly modified. Among the different histone modifications that can occur (acetylation, methylation, ubiquitylation, phosphorylation, etc.), histone acetylation and deacetylation are best understood.

a. **Histone modification by acetylation and deacetylation of lysine residues:** These processes are important in making DNA more or less accessible to transcription factors (proteins that regulate gene expression by direct binding to DNA). Lysine residue acetylation weakens the DNA-histone interactions and makes the DNA more accessible to factors needed for transcrip-

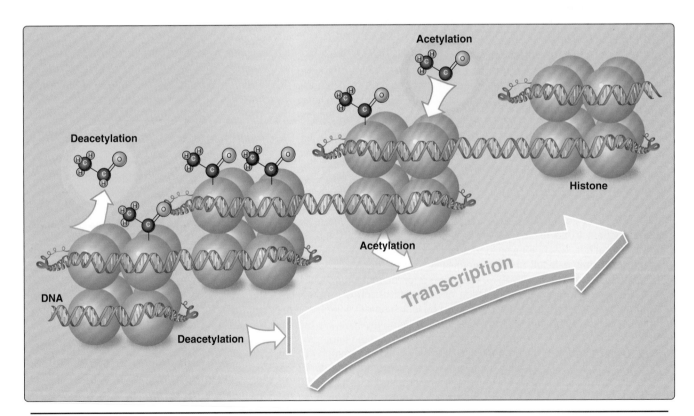

Figure 6.12
Histone (de)acetylation controls chromatin compaction and decompaction.

tion (Figure 6.12). Therefore, **histone acetylation** (catalyzed by **h**istone **a**cetyl**t**ransferases or **HAT**s) is generally associated with **transcriptional activation**. On the other hand, **histone deacetylation** (catalyzed by **h**istone **deac**etylase or **HDAC**) is associated with gene **silencing**. The interplay of acetylase and deacetylase activities defines the transcriptional activity of a given chromatin region.

> ## Histone code, epigenetic readers, writers, and erasers
>
> Cumulative evidence from studies of the various histone modifications suggests that the transcription of genetic information in DNA is partly regulated by specific types of modifications. The histone code, for histone methylation, can be associated with either transcriptional activation or repression. For example, trimethylation of histone H3 at lysine 4 (H3K4me3) is an active mark for transcription, whereas trimethylation of histone H3 at lysine 27 is an active mark of repression. Histone acetyltransferases and histone methyltransferases are enzymes that can "write" the code, while histone deacetylases and histone demethylases can "erase" the code. Yet another class of proteins have been discovered with specific conserved domains that recognize modified histones to "read" and interpret the histone modification language.

b. **Chromatin remodeling complexes:** It is now recognized that chromatin remodeling complexes work in concert with histone modifications to regulate gene expression by helping to move, reposition, or eject nucleosomes thus creating a nucleosome-free region to facilitate the binding of transcription factors. While several families of chromatin remodelers are now known in eukaryotes, SWI/SNF (switch-sniff, the first remodeling complex discovered) is better understood and uses ATP as an energy source for disruption of the many contacts between DNA and nucleosomes for gene activation.

Chapter Summary

- Chromatin is composed of DNA and small histone proteins.
- DNA packaging requires histone's amino acid side chains to interact with DNA.
- Histone modifications are reversible, thus enabling chromatin compaction and decompaction.
- Telomeres, centromeres, and multiple origins of replication are important for chromosome maintenance and replication.
- Genomic DNA contains both repeat and unique sequences.
- Different cells express different regions of chromatin.
- Methylation and tissue-specific chromatin structure represent stably inherited epigenetic changes in the genome.
- Methylation of certain cytosines in DNA correlates with transcriptional silencing.

Study Questions

Choose the ONE best answer.

6.1 Telomeres

    A. Consist of repetitive DNA sequences found at the ends of chromosomes.

    B. Inhibit the organization of the DNA as nucleosome units in chromosomes.

    C. Facilitate the attachment of chromosomes to kinetochore during cell division.

    D. Are the regions of chromosomal DNA where gene clusters are located.

    E. Are interspersed throughout the chromosomes with unique DNA sequences.

> Correct answer = A. Telomeres are repeat sequences found at the ends of the chromosomes that protect them from damage. They do not affect the organization of DNA into higher-order structures. The kinetochore is formed around the centromere region. The ends of DNA do not contain genes. LINES and SINES are the moderately repeated sequences that are interspersed with unique DNA sequences.

6.2 Which of the following statements regarding histone proteins is CORRECT?

    A. Histone proteins contain large amounts of acidic amino acid residues.

    B. Histone proteins are important to stabilize the single-stranded DNA.

    C. Histones constitute three times the mass of DNA within the nucleus.

    D. Nuclear DNA associates with histone proteins to form chromatin.

    E. Within chromosomes, each nucleosome unit contains a different histone protein.

> Correct answer = D. The term chromatin refers to the genetic material found in the nucleus complexed to histones. Histones are basic proteins containing a large amount of basic amino acids. Double-stranded DNA is wrapped around histone beads. There is an equal mass of DNA to histones. Nucleosomes are the fundamental level of organization containing the same histone proteins.

6.3 Tandemly repeated DNA sequences form a part of

    A. Single-copy nuclear DNA.

    B. Mitochondrial DNA.

    C. Long interspersed elements (LINES).

    D. Minisatellite DNA.

    E. Short interspersed elements (SINES).

> Correct answer = D. Tandemly repeated sequences form a part of the minisatellite DNA sequence present in the nuclear DNA. Single-copy genes contain unique DNA sequences. LINES and SINES are moderately repetitive DNA sequences, which are interspersed throughout the nuclear genome.

6.4 In which of the following cells/tissues would you expect the β-globin gene to be unmethylated?

    A. Erythroid cells

    B. Kidney

    C. Liver

    D. Skin

    E. White blood cells

> Correct answer = A. The globin gene is unmethylated and active in erythroid cells, which synthesize hemoglobin. All other cell types do not require globulin protein synthesis and are, therefore, methylated and silenced.

6.5 Modification of histone proteins by acetylation will

    A. Add methyl groups to the regulatory region of the target genes.

    B. Increase the condensation of chromatin.

    C. Increase the affinity of histones for DNA.

    D. Increase the transcription of target genes.

    E. Inhibit RNA polymerase activity.

> Correct answer = D. Modification of histones by acetylation decreases the affinity for DNA and causes decompaction of the chromatin, allowing gene transcription to take place. Histones do not control the addition of methyl groups to DNA. Deacetylation of histones increases their affinity to DNA. Modification of histones by acetylation will generally result in activating RNA polymerase to bind DNA and initiate transcription.

DNA Replication

7

I. OVERVIEW

Deoxyribonucleic acid (DNA) contains all the information necessary for the development and function of all organisms. The replication or copying of cellular DNA occurs during the S or synthesis phase of the cell cycle (see Chapter 20). This is a necessary process to ensure that the instructions in DNA are faithfully passed on to the newly produced cells. In cell nuclei, DNA and protein complexes known as **chromatin** make up the **chromosomes** (see Chapter 6). Because replication can occur only on a single-stranded DNA template, the double-stranded DNA of the chromatin must first unwind. Once unwound, both strands of DNA are copied simultaneously. This process requires proteins to break open the double-stranded DNA, forming a **replication fork**. The main enzyme that catalyzes the formation of new DNA strands is **DNA polymerase**. Faithful copying of DNA requires that the DNA polymerase recognize nucleotides on the opposite strand of the DNA and has a proofreading function to recognize and correct any mistakes that have been made. DNA that is being replicated experiences **torsion**, or twisting, created during the unwinding of DNA. Enzymes called **topoisomerases** act to reduce this torsional force. (Topoisomerases are an important pharmacological target for drug agents designed to inhibit DNA replication.) After replication is complete, the parent and the daughter strands of DNA must re-form a double-stranded structure as well as reestablish the chromatin structure. This area of eukaryotic molecular biology is one in which gaps in our knowledge exist. The steps and processes known to occur in the replication of prokaryotic (bacterial) DNA generally apply to eukaryotic DNA replication as well.

II. DNA STRUCTURE

The structure of DNA was first described by James Watson and Francis Crick in 1953. DNA exists as a **double helix**, with about 10 nucleotide pairs per helical turn. The spatial relationship between the two strands creates furrows in DNA—the **major and minor grooves**. Each of the two helical strands is composed of a sugar phosphate backbone with attached bases and is connected to a complementary strand by hydrogen bonding. The sugar in DNA is **deoxyribose**. The pairing of the nucleotide bases occurs such that adenine (A) binds with thymine (T) and guanine (G) with cytosine (C).

A. Primary structure

The order of the nucleotide bases determines the primary structure or sequence of the DNA. Overall, DNA is a high molecular weight double-stranded polymer (>10^8) of deoxyribonucleotides joined by covalent **phosphodiester bonds**. The phosphodiester bonds are bonds that form between the 3′-OH groups of the deoxyribose sugar on one nucleotide with the 5′ phosphate groups on the adjacent nucleotide (Figure 7.1). The phosphodiester linkages between individual deoxynucleotides are directional in nature. The 5′ phosphate group of one nucleotide is bound to the 3′ hydroxyl group of the next nucleotide. The two complementary strands of DNA double helix therefore run in **antiparallel** directions. The 5′ end of one strand is base-paired with the 3′ end of the other strand. This primary structure is stabilized by two types of noncovalent interactions (Figure 7.2).

B. Noncovalent interactions

One type of noncovalent interaction within DNA includes **hydrogen bonds** that hold together the two strands of DNA within the double

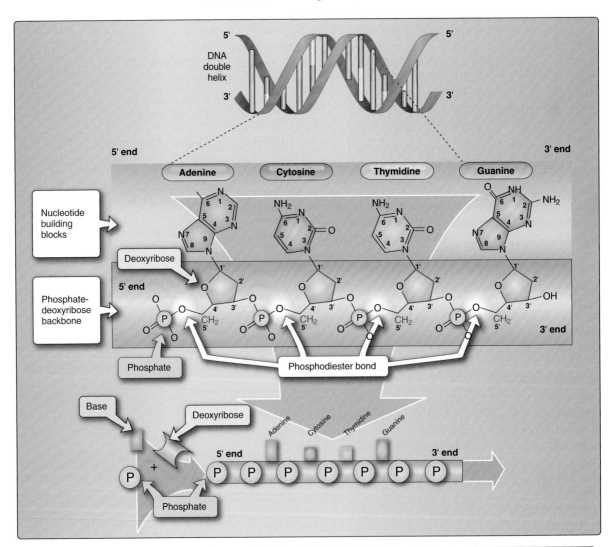

Figure 7.1
The covalent structure of DNA.

helical structure. Nucleotide bases on one strand form these bonds with nucleotide bases on the opposite strand. Adenine forms two hydrogen bonds with thymine, while guanine and cytosine are connected by three hydrogen bonds. This type of base pairing in the interior of the helix stabilizes the interior of the double-stranded DNA because the stacked bases repel each other due to their **hydrophobic** nature. The hydrogen bonds between bases can be made and broken easily, allowing DNA to undergo accurate replication and repair (Figure 7.2).

III. CHARACTERISTICS OF EUKARYOTIC DNA SYNTHESIS

During the process of eukaryotic DNA replication, the enzymes known as DNA polymerases select the nucleotide that is to be added to the 3'-OH end of the growing chain and catalyze the formation of the phosphodiester bond. The substrates for DNA polymerases are the four deoxynucleoside triphosphates (dATP, dCTP, dGTP, and dTTP) and a single-stranded template DNA. There are distinct differences in the mechanism of DNA replication between prokaryotic and eukaryotic DNA. The focus here is on the eukaryotic processes, and the following are some of the characteristic features of the eukaryotic DNA replication.

A. Semiconservative with respect to parental strand

One characteristic of eukaryotic DNA replication is that it is a semiconservative process. When DNA is replicated during the process of cell division, one parent or original strand of DNA is distributed to each daughter duplex in combination with a newly synthesized strand with an antiparallel orientation. Because genetic information on both strands is similar, at the end of the process, each of the two daughter strands has half new DNA and half old DNA; thus, the process is semiconservative (Figure 7.3).

B. Bidirectional with multiple origins of replication

Another characteristic of eukaryotic DNA replication is that it is bidirectional and starts in several different locations at once. DNA is copied at about 50 base pairs (bp) per second. For eukaryotic DNA consisting of 3×10^9 nucleotides, this process would take an extremely long time rather than the hours it actually does. This hastening of the replication process is made possible by the fact that replication begins at several sites on linear DNA and is completed by the end of S phase of the cell cycle (Chapter 20). As replication nears completion, "bubbles" of newly replicated DNA come together forming two new molecules (Figure 7.4).

C. Primed by short stretches of RNA

A third characteristic of eukaryotic DNA replication is that it requires a short stretch of ribonucleic acid (RNA) for the initiation of the process. DNA polymerases cannot initiate synthesis of a complementary strand of DNA on a totally single-stranded template. A specific DNA polymerase–associated enzyme, called **DNA primase**, synthesizes short stretches of RNA that are complementary and antiparallel to the DNA template. The RNA primer is later removed. Chain elongation is carried out by DNA polymerases by the addition of deoxyribonucleotides to

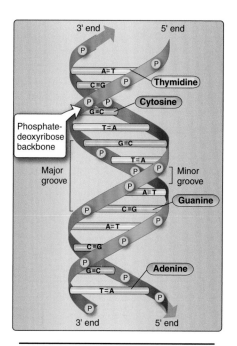

Figure 7.2
DNA double helix.

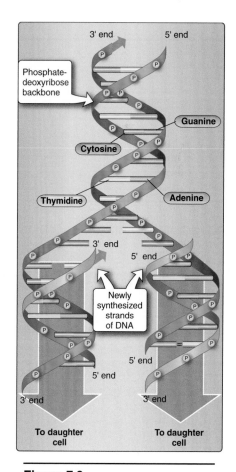

Figure 7.3
DNA synthesis is semiconservative with respect to the parental strand.

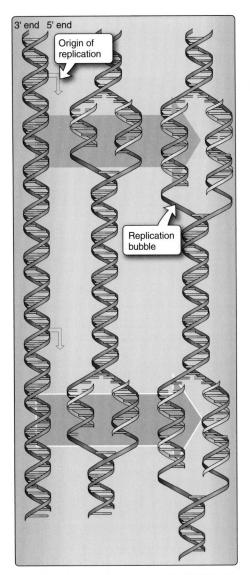

Figure 7.4
Bidirectional with multiple origins of replication.

the 3′ end of the growing chain. The sequence of nucleotides that are added is dictated by the base sequence of the template (or coding) strand with which the incoming nucleotides are paired (Figure 7.5).

D. Semidiscontinuous with respect to the synthesis of new DNA

An additional characteristic of eukaryotic DNA replication is that it is a semidiscontinuous process. A new strand of DNA is always synthesized in the 5′ to 3′ direction. Because the two strands of DNA are antiparallel, the strand being copied is read from the 3′ end toward the 5′ end.

All DNA polymerases function in the same manner: They "read" a parental strand 3′ to 5′ and synthesize a complementary antiparallel new strand 5′ to 3′.

Because parental DNA has two antiparallel strands, the DNA polymerase synthesizes one strand in the 5′ to 3′ continuously. This strand is called the **leading strand**. The continuous or leading strand is the one in which 5′ to 3′ synthesis proceeds in the same direction as replication fork movement. The other new strand is synthesized 5′ to 3′, but discontinuously, creating fragments that ligate (join) together later. This strand is called the discontinuous or **lagging strand** and is the one in which 5′ to 3′ synthesis proceeds in the direction opposite to the direction of the fork movement. The DNA synthesized on the lagging strand as short fragments (100 to 200 nucleotides) is called the **Okazaki fragments**. Although overall chain growth occurs at the base of the replication fork, synthesis of the lagging strand occurs discontinuously in the opposite direction but with exclusive 5′ to 3′ polarity (Figure 7.6).

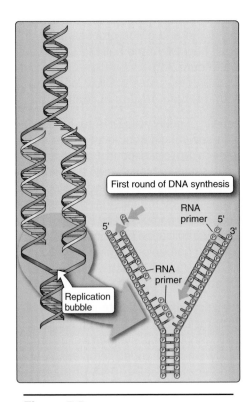

Figure 7.5
Primed by short stretches of RNA.

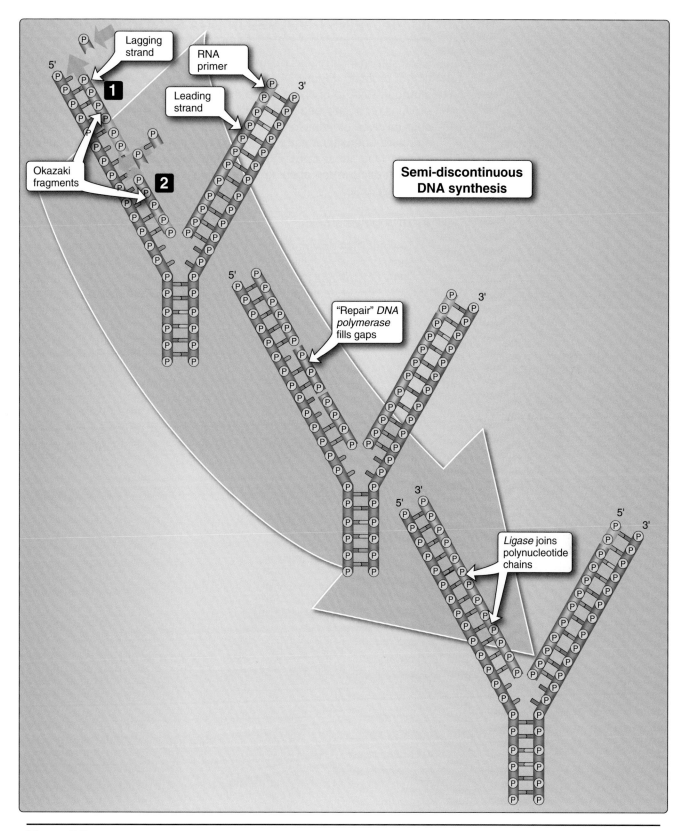

Figure 7.6
Semidiscontinuous with respect to the synthesis of new DNA. (Not to scale.)

IV. PROTEINS INVOLVED IN DNA SYNTHESIS

Each step in the process of eukaryotic DNA replication requires the function of proteins (Figure 7.7). For example, it is important for the double-stranded DNA to open up at the multiple origins of replication and have proteins recognize and bind to DNA and prime it for replication. This process involves proteins to break the double-stranded DNA, keep the DNA structure open, synthesize a new strand, and, finally, put them together as one long linear DNA. Other proteins are needed to remove torsion that can occur when opening a double helix. Some of these proteins are described below.

A. DNA polymerases

Several DNA polymerases are involved in DNA replication and each of them possesses distinct activities (Table 7.1). The individual eukaryotic DNA polymerases have particular associated activities. They function as a complex to initiate DNA synthesis. Some DNA polymerases have 3′ to 5′ exonuclease activity, or **proofreading** ability, that allows them

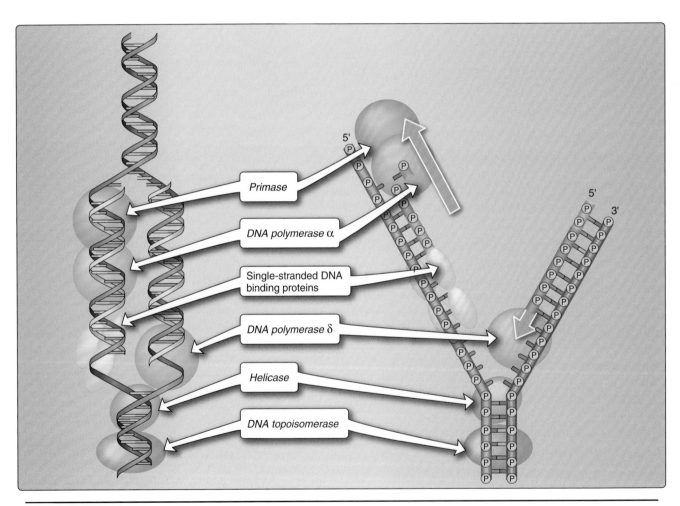

Figure 7.7
Proteins involved in DNA synthesis.

Table 7.1: Properties of Eukaryotic DNA Polymerases

Polymerase	α	β	γ	δ	ε
Location	Nucleus	Nucleus	Mitochondria	Nucleus	Nucleus
Replication	Yes	No	Yes	No	Yes
Repair	No	Yes	No	No	Yes
Associated functions: 5′–3′ polymerase	Yes	Yes	Yes	Yes	Yes
3′–5′ exonuclease	No	No	Yes	Yes	Yes
5′–3′ exonuclease	No	No	No	No	No

to remove nucleotides that are not part of the double helix. The enzyme removes mismatched residues, thus performing an editing function. This activity enhances the fidelity of DNA replication by rechecking the correctness of base pairing before proceeding with polymerization (Figure 7.8). Eukaryotic polymerases *do not* possess 5′ to 3′ exonuclease activity, an activity that is important for removing primers in prokaryotes. In eukaryotes, primer removal is done by another enzyme.

B. DNA helicases

DNA helicases are a class of motor proteins required to unwind short segments of the parental duplex DNA. These enzymes use energy generated from nucleotide (adenosine triphosphate [ATP]) hydrolysis to catalyze the strand separation and formation of the replication fork during DNA synthesis.

C. DNA primases

DNA primases initiate the synthesis of an RNA molecule essential for priming DNA synthesis on both the leading and the lagging strands. The first few nucleotides are ribonucleotides, and the subsequent ones may be either ribonucleotides or deoxyribonucleotides.

D. Single-stranded DNA-binding proteins

Single-stranded DNA-binding proteins prevent premature annealing of the single-stranded DNA to double-stranded DNA. An important function of single-stranded DNA proteins during the process of DNA replication is to keep the strands protected until the complementary strands are produced (Figure 7.7).

E. DNA ligase

DNA ligase is an enzyme that catalyzes the sealing of nicks (breaks) remaining in the DNA after DNA polymerase fills the gaps left by RNA primers. DNA ligase is required to create the final phosphodiester bond between the adjacent nucleotides on a strand of DNA (Figure 7.9).

F. Topoisomerases

Most cellular DNAs have fewer right-hand turns than expected from the number of their base pairs. This underwound condition (negative **supercoils**) facilitates the unwinding of the double helix during replication and transcription. As the replication fork moves along the

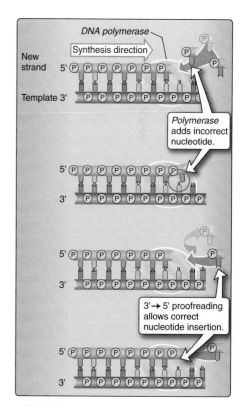

Figure 7.8
Proofreading activity of some DNA polymerases.

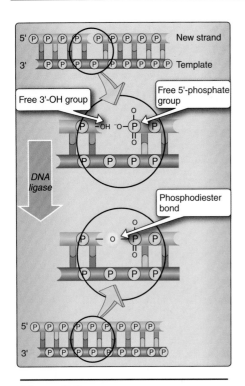

Figure 7.9
Mechanism of action of DNA ligase.

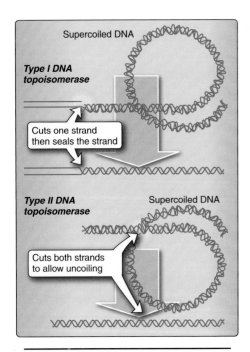

Figure 7.10
Mechanism of action of topoisomerases.

helix, rotation of the daughter molecules around one another causes the DNA strands to become overwound. The supertwisting of DNA can be removed by enzymes known collectively as topoisomerases. These enzymes relieve torsional stress in DNA by inducing reversible single-stranded breaks in DNA. The phosphodiester bond either in one strand or in both strands is cleaved initially. After rotation of the DNA around its axis, the enzyme seals the nick.

1. **Topoisomerase I:** This form of topoisomerase catalyzes breaks in only one strand of the double-stranded DNA, allowing unwinding of the broken strand and then rejoining of the broken ends by catalyzing the formation of new phosphodiester bonds (Figure 7.10).

2. **Topoisomerase II:** This form of topoisomerase catalyzes breaks in both strands of the double-stranded DNA, allowing both broken strands to unwind, and then catalyzes the formation of new phosphodiester bonds (see Figure 7.10).

Clinical Application 7.1: Topoisomerase Activities as Targets for Antibiotics

DNA gyrase is a bacterial type II topoisomerase that functions ahead of the replication fork. It relaxes DNA molecules in an ATP-dependent fashion. **Nalidixic acid** and **norfloxacin** are drugs used in the treatment of urinary tract and other infections. These compounds inhibit bacterial DNA gyrase by inhibiting the strand-cutting reaction. Human type II topoisomerase is much less sensitive to the action of these two drugs. **Doxorubicin**, **etoposide**, and **teniposide** inhibit human topoisomerase II and are used in the treatment of several neoplastic diseases (cancers). These drugs act by enhancing the rate at which target topoisomerase II cleaves DNA and by reducing the rate at which these breaks are resealed.

G. Telomerase

Telomerase is an enzyme that helps to maintain the telomere. The telomere, a protective repetitive stretch of DNA complexed with protein at the end of a chromosome, shortens with every cell division. Telomere shortening is recognized as and is a part of the normal aging process. Telomeres are important chromosomal structures and allow the cell to distinguish intact chromosomes from broken chromosomes and to protect chromosomes from degradation. They also serve as substrates for normal replication mechanisms. In most organisms, telomeric DNA consists of a tandem array of very simple sequence of DNA (in humans, it is TTAGGG).

The telomere maintenance enzyme, telomerase, is an RNA-dependent DNA polymerase, which adds TTAGGG repeats to the ends of the chromosomes. The telomerase ribonucleoprotein complex contains an RNA template, which is an integral component of the enzyme. With the help of its RNA template, it adds a series of DNA repeats to the leading strand. This addition allows the lagging strand to be completed by DNA polymerase (Figure 7.11). Some normal cells (normally regenerating tissues, stem cells, and progenitor cells) express telomerase. Intact telomere function is required for tissue homeostasis. Cancer cells appear to reactivate this enzyme, which circumvents the problem of end replication and immortalizes them.

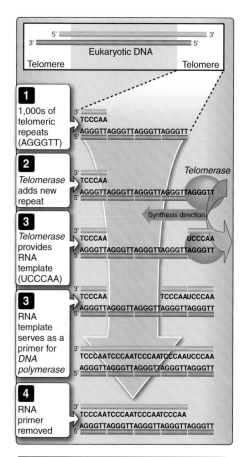

Figure 7.11

Mechanism of action of telomerase.

Clinical Application 7.2: Cloned Sheep, Dolly, Shows her Age

Dolly was the first animal to be cloned in 1997, using somatic cells and the process of nuclear transfer. The somatic cells were obtained from the mammary gland of a donor sheep. Dolly, then, is a genetic duplicate of the donor sheep. While she appeared normal at birth and had normal development until age 3, chromosomal examination revealed that her telomeres were shorter than would have been expected for a sheep of her chronological age. In fact, her telomeres were found to be the average length of those expected in a 6-year-old sheep—the age of the ewe from whose cells Dolly was cloned. Dolly died of lung disease at 6 years of age, leading some scientists to believe her biological age was indeed much older than her chronological age.

V. DNA DAMAGE

DNA damage can result from both endogenous and exogenous causes. Most DNA damage is repaired before DNA is replicated. Mutagenic agents (ones that induce mutations) are therefore most effective in causing their damage during S phase of the cell cycle when the new DNA is being synthesized.

A. Basal mutation rate

The rate of mutations occurring from endogenous (internal cellular) causes is termed the basal mutation rate. This is the mutation rate observed in the absence of environmental mutagens and is caused by errors during DNA replication. Spontaneous tautomeric shifts (changes from one natural structural form to another) in the bases contribute to these errors. Fortunately, these bases spend very little time in their less stable forms, so mutations caused by tautomeric shift are rare (Figure 7.12).

B. Exogenous agents

Outside influences can also affect the mutation rate of DNA. For example, **ionizing radiation,** including X-rays and radioactive radiation, is sufficiently energy rich to react with DNA. Ionizing radiation penetrates the whole body and, therefore, can cause both somatic (cellular) and germ line (oocyte or sperm) mutations. (Ultraviolet radiation is nonionizing and cannot penetrate beyond the outer layers of the skin. Nevertheless, ultraviolet radiation from sunlight can be mutagenic [see Nucleotide excision repair below].) Some chemicals in the environment, including **hydrocarbon**s, are known to induce mutations. Hydrocarbons found in cigarette smoke are well-known mutagens. **Oxidative free radicals**, produced from internal

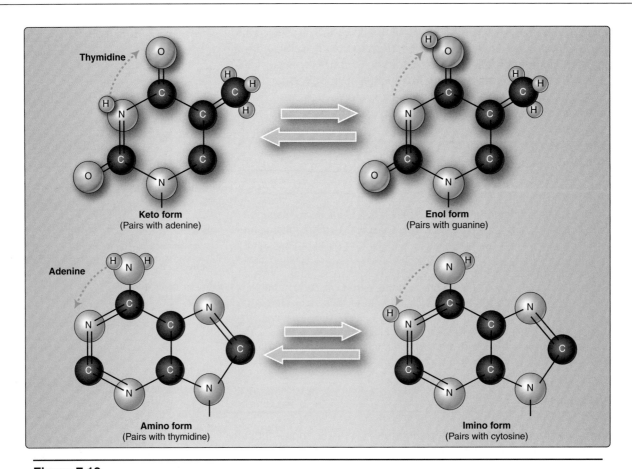

Figure 7.12
Bases in DNA undergo tautomeric shifts.

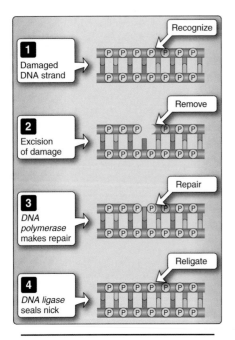

Figure 7.13
General scheme of action of DNA repair systems.

or external sources, can also induce DNA damage. Chemicals used in **chemotherapy**, especially for the treatment of cancer, can also induce mutations.

VI. DNA REPAIR SYSTEMS

DNA repair is necessary not only because cells are continuously exposed to environmental mutagens but also because thousands of mutations would otherwise occur spontaneously in every cell each day during DNA replication. A variety of strategies exist for repairing damage to DNA. In most of these cases, the cells use the undamaged strand of DNA as a template to correct the mistakes in DNA. When both strands are damaged, the cell resorts to the use of the sister chromatid (the second copy of DNA present in diploid cells) or to an error-prone recovery mechanism. When defects in DNA repair mechanisms occur, mutations accumulate in the cell's DNA leading to cancer. All types of repair mechanisms are made up of enzymes that follow a general scheme of recognition, removal, repair, and religation. However, depending on the type of damage, different enzymes are employed (Figure 7.13).

A. Mismatch repair

Mismatch repair deals with correcting the mismatches of normal bases that fail to maintain normal Watson-Crick base pairing (A to T, C to G) and insertions and deletions of one or a few nucleotides that are introduced into DNA during replication. This failure is typically due to the mistakes made by DNA polymerase during replication. In eukaryotes, recognition of a mismatch is accomplished by several different proteins including those encoded by *MSH2*, *MLH1*, *MSH6*, *PMS1*, and *PMS2* genes (Figure 7.14). Mutations in either of these genes predispose the person to an inherited form of colon cancer (**hereditary nonpolyposis colon cancer [HNPCC]**) at a young age. Other cancers (endometrial, ovarian, stomach, etc.) are known to occur in the affected families.

B. Base excision repair

Base excision repair is required to correct the spontaneous depurination and spontaneous deamination (removal of amine groups) that happen to bases present in DNA. About 10,000 purine (adenine and guanine) bases are lost per cell per day. Spontaneous deamination of cytosine causes it to be converted to uracil, normally found in RNA and not in DNA. Methyl cytosine in DNA (see Chapter 6) is converted to thymine on spontaneous deamination resulting in the most common mutation seen in humans a C to a T transition. Base excision repair involves recognition and removal of nucleotides that have lost the bases or have been modified (Figure 7.15).

C. Nucleotide excision repair

This type of repair is necessary to remove ultraviolet light–induced DNA damage as well as DNA damage from environmental chemicals. UV light is nonionizing and cannot penetrate beyond the outer layer of the skin, but it can nevertheless form **pyrimidine-pyrimidine dimers** (commonly thymine-thymine dimers) from adjacent pyrimidine bases (cytosine and guanine) in DNA. Thus, sunlight is mutagenic, causing both sunburn and skin cancer (Figure 7.16).

This repair mechanism is also necessary to recognize chemically induced bulky additions to DNA that like thymine-thymine dimers distort the shape of the DNA double helix and cause mutations. Carcinogens, such as benzopyrene in cigarette smoke, react with DNA, causing mutations. The enzymes involved in this repair pathway consist of several proteins (about 30) that are required for the nucleotide excision repair process.

Clinical Application 7.3: Xeroderma Pigmentosum

Xeroderma pigmentosum is a genetic disorder of DNA repair in which patients carry mutations in the nucleotide repair enzymes. Analysis of humans with xeroderma pigmentosum suggests that several proteins may be required for the excision of damaged bases in DNA by a single repair system. The condition is inherited in an autosomal recessive fashion and is characterized by defective repair of thymine dimers. Affected individuals are prone to develop multiple skin cancers. The reduced ability to repair DNA leads to somatic mutations, some of which result in malignant transformation (Figure 7.17).

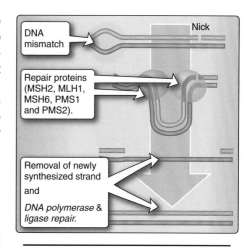

Figure 7.14
Mismatch repair.

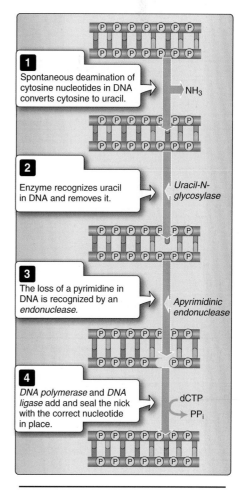

Figure 7.15
Base excision repair.

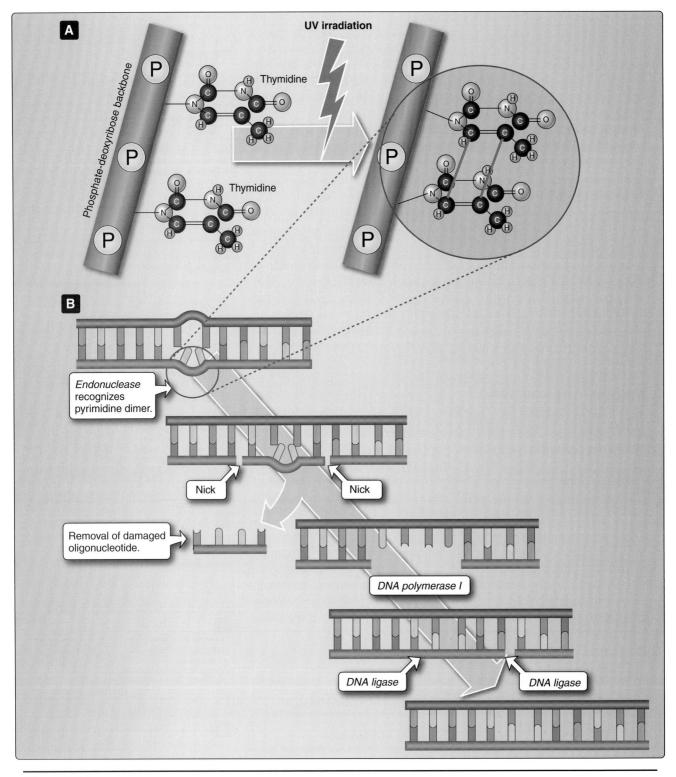

Figure 7.16
A. Pyrimidine-pyrimidine dimer formation in DNA. **B.** Nucleotide excision repair.

D. Double-stranded DNA repair

When damage from ionizing radiation, oxidative free radicals, or chemotherapeutic agents causes both the strands of DNA to be severed, two types of repair mechanisms exist to correct the damage, homologous recombination and nonhomologous end joining (Figure 7.18).

1. **Homologous recombination:** This type of repair takes advantage of sequence information available from the unaffected homologous chromosome for proper repair of breaks. **BRCA1 and BRCA2** proteins normally play a role in the homologous recombination process. Mutations in these genes increase breast cancer risk. Fanconi anemia is a condition caused by failures in DNA recombination repair enzymes to correct the defects by homologous recombination. Several Fanconi anemia proteins form complexes and interact with the BRCA proteins.

2. **Nonhomologous end joining:** This process permits the joining of ends even if there is no sequence similarity between them. It is error prone since it can also introduce mutations during repair. Nonhomologous end joining is especially important before the cell has replicated its DNA because there is no template available for repair by homologous recombination.

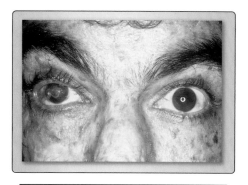

Figure 7.17
Patient with xeroderma pigmentosum.

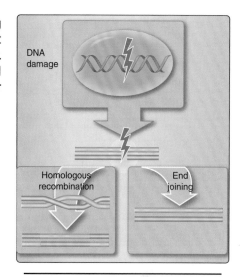

Figure 7.18
Double-stranded DNA break repair.

Chapter Summary

- Eukaryotic DNA replication is bidirectional, is semiconservative, requires a primer, and can only occur in the 5′ to 3′ direction.
- Several proteins are required for DNA synthesis; there are differences between the prokaryotic and eukaryotic enzymes that function in DNA synthesis.
- DNA polymerase can "proofread," which is made possible by its 3′ to 5′ exonuclease activity. This reduces the copying errors made by DNA polymerase.
- Topoisomerases are required to remove torsional stress in DNA. Several drugs are the targets of topoisomerases.
- Telomerase is an RNA-dependent DNA polymerase and can extend telomeres. This activity, however, is not present in normal diploid cells.
- Basal mutation rate refers to mistakes made endogenously in the cell and is usually a result of errors by DNA polymerase during replication.

Chapter Summary (Continued)

- A number of environmental agents can cause mutations in DNA—UV light and other ionizing radiations, chemicals, and chemotherapeutic agents.
- Several types of repair mechanisms exist to correct the mistakes in DNA structure.
- Most of the repair mechanisms depend on the presence of an intact complementary DNA sequence.
- Double-stranded DNA breaks are repaired by two different processes, one requiring a homologous chromosome and the other error-prone mechanism of nonhomologous end joining.

Study Questions

Choose the ONE best answer.

7.1 During eukaryotic DNA replication, topoisomerases

   A. Catalyze the synthesis of an RNA primer on the lagging strand.

   B. Remove the incorrectly base-paired nucleotides through a 3′ to 5′ exonuclease activity.

   C. Stabilize single-stranded DNA in the region of the replication fork.

   D. Cut and reseal DNA in advance of the replication fork to eliminate supercoiling.

   E. Add nucleotides to the growing strand in the 5′ to 3′ direction.

Correct answer = D. Topoisomerases remove torsion from replicating DNA by cutting the double-stranded DNA, relaxing the supercoiling followed by ligation. DNA primase catalyzes the addition of an RNA primer during DNA synthesis. Proofreading activity is a property of some DNA polymerases. Single-stranded DNA-binding proteins protect the DNA during replication by binding to the open template strand. DNA polymerase synthesizes DNA in the 5′ to 3′ direction by adding nucleotides to the growing chain.

7.2 Which of the following functions is associated with eukaryotic DNA polymerase during DNA replication?

   A. Continuous 5′ to 3′ DNA synthesis on the lagging strand

   B. Energy-dependent formation of the replication fork

   C. Proofreading newly synthesized DNA

   D. Discontinuous 5′ to 3′ DNA synthesis on the leading strand

   E. Removal of primers using 5′ to 3′ exonuclease activity

Correct answer = C. Some of the DNA polymerases possess proofreading ability, which is a 3′ to 5′ exonuclease activity. Continuous DNA synthesis occurs on the leading strand and discontinuous synthesis on the lagging strand. Helicase is necessary for breaking the hydrogen bonds between the strands of DNA using ATP hydrolysis. None of the DNA polymerases possess 5′ to 3′ exonuclease activity.

7.3 Defective DNA mismatch repair can result in the development of

   A. Hereditary nonpolyposis colorectal cancer.

   B. Skin cancer.

   C. Sunburns.

   D. UV light–induced damage.

   E. Xeroderma pigmentosum.

Correct answer = A. Loss of mismatch repair activity predisposes individuals to a form of colon cancer called hereditary nonpolyposis colon cancer. Loss of nucleotide excision repair results in a greater susceptibility to sunburns and skin cancer and an inability to repair UV light–induced pyrimidine-pyrimidine dimer formation. A loss of function of any of the enzymes involved in nucleotide excision repair predisposes individuals to the syndrome xeroderma pigmentosum.

7.4 A 6-year-old male patient presents with photosensitivity and multiple skin tumors. Which of the following types of DNA damage most likely accounts for his condition?

A. Double-strand DNA damage
B. Deaminated cytosines
C. Mismatched base pairs
D. Thymine dimers
E. Depurinated DNA

Correct answer = D. Thymine-thymine dimers are the most common type of sunlight-induced DNA damage that is repaired by the nucleotide excision system. Double-stranded DNA break is repaired by one of two systems that utilize homologous chromosomes or cause nonhomologous end joining. Deaminated cytosines and depurinated DNA are fixed by the base excision repair system. Mismatches occur in the DNA due to errors made by the DNA polymerase during replication.

7.5 Treatment of cycling cells with a topoisomerase II inhibitor, such as doxorubicin, will result directly in

A. A decrease in the time needed for DNA to replicate.
B. A decreased number of errors during replication of DNA.
C. Lengthening of the ends of chromosomes.
D. Removal of torsion from replicating DNA.
E. The cleavage of replicating DNA into smaller fragments.

Correct answer = E. Doxorubicin inhibits the ligase-like action of topoisomerase II. Therefore, DNA will be initially cleaved and relaxed but will not be ligated back, resulting in the fragmentation of DNA. The action of this drug will slow down DNA replication and increase errors in the replicating DNA. Telomerase lengthens the ends of chromosomes, while inhibition of topoisomerases will cause more torsion in DNA.

8

Transcription

I. OVERVIEW

Transcription refers to the first step in gene expression, the copying of a particular sequence of deoxyribonucleic acid (DNA) into messenger ribonucleic acid (mRNA). Genes are considered **expressed** when the information contained within DNA has been converted to proteins that affect cellular properties and activities. The DNA-directed synthesis of mRNA is a necessary intermediate to produce protein.

In order for an mRNA to be produced, a gene sequence on DNA needs to be identified, along with information necessary for the exact start site. Genes are split into exons and introns, and the entire region is initially transcribed. This primary RNA transcript is processed before it exits the nucleus. Once created, mRNA is modified through RNA splicing, 5′ end capping, and the addition of a poly(A) tail after which the mature mRNA enters the cytoplasm.

Generally, every gene contains two classes of information, one to specify the primary structure of the final product and the other to regulate the expression of the gene. Both the timing and the amount of RNA produced are regulated during transcription. mRNA encodes the amino acid sequence of proteins, and both ribosomal and transfer RNAs directly participate in protein synthesis.

II. TYPES OF RNA

Several distinct types of RNA are known: ribosomal (rRNA), transfer (tRNA), messenger (mRNA), and other small noncoding RNA, each with its own distinctive structure and function. These RNAs (tRNA and rRNA) are transcribed but not translated and are considered noncoding RNAs (ncRNAs). There are additional small RNAs that are also noncoding, like the ones in the nucleolus (snoRNAs) nucleus (snRNA) and cytoplasm (miRNA) that perform specialized functions.

A. Ribosomal RNA

rRNA accounts for approximately 80% of total RNA in the cell and associates with proteins to form ribosomes. Eukaryotes have several different rRNA molecules: 18S, 28S, 5S, 5.8S, 18S, and 28S. Ribosomes are important during protein synthesis as they contain peptidyltransferase "activity," an activity catalyzed by ribozymes (Figure 8.1).

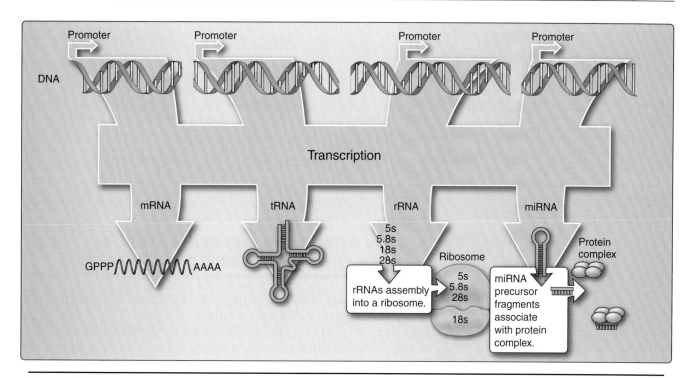

Figure 8.1
Different types of eukaryotic RNA.

B. Transfer RNA

tRNA is the smallest of the three RNAs. It functions in the protein synthesis by virtue of its ability to carry the appropriate amino acid and also provide a mechanism by which nucleotide information can be translated to amino acid information through its anticodon.

C. Messenger RNA

mRNA carries genetic information from DNA to cytosol for translation. About 5% of the total RNA within a cell is mRNA. It is the most heterogeneous in terms of size and carries specific information necessary for the synthesis of different proteins.

D. MicroRNA

miRNAs, like the other RNA molecules, are encoded by genes and are single-stranded RNA molecules about 21 to 23 nucleotides in length. These newly discovered molecules are transcribed but not translated. They function in regulating gene expression by their ability to bind mRNA and to down-regulate the gene expression.

Distinct enzymes catalyze the synthesis of various RNAs as shown in Table 8.1.

Table 8.1: Eukaryotic RNA Polymerases

Polymerase	RNA Products
RNA polymerase I	Ribosomal RNAs
RNA polymerase II	Messenger RNA, microRNA, and some noncoding RNAs
RNA polymerase III	Transfer RNAs

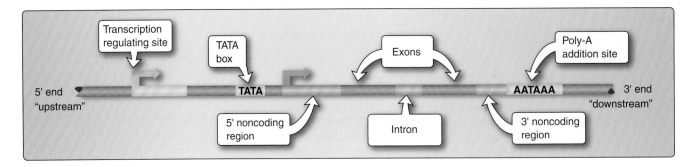

Figure 8.2
Structure of a typical eukaryotic gene.

III. GENE STRUCTURE AND REGULATORY ELEMENTS IN EUKARYOTIC PROTEIN-CODING GENES

The minimal linear sequence of genomic nucleic acids that encode proteins and structural RNA is termed a **gene** (Figure 8.2). Gene sequences are written from 5′ (5 prime) to 3′. Eukaryotic genes are composed of coding exons, noncoding introns, and noncoding consensus sequences. The number of introns and exons, their size, location, and sequence differ from gene to gene. Noncoding regions at the 5′ end to the first exon are referred to as upstream sequences and those at the 3′ end are called downstream sequences.

A. Consensus sequences

Consensus sequences are evolutionarily conserved and act as recognition markers and define a potential DNA recognition site. They are usually bound by proteins (**transcription factors)** and other regulatory proteins that recognize a particular sequence.

1. **Promoters:** Promoters are DNA sequences that select or determine the start site of RNA synthesis. The consensus sequence for promoters typically has the sequence "TATA" (or variations of T and A) and is often located 15 to 30 base pairs (bp) upstream from the transcription start site, called a TATA box and an initiator sequence (Inr) near the RNA start site at +1 (Figure 8.3). Additional sequences that may be required for promoter function include the CAAT box

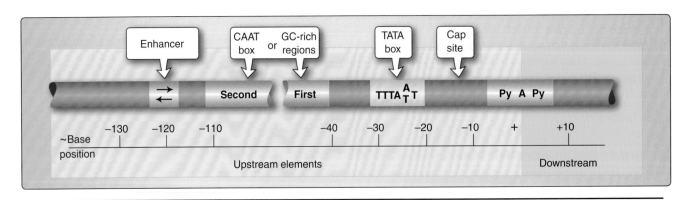

Figure 8.3
Promoter elements found upstream to the coding sequences in a gene.

and the GC box. In eukaryotes, proteins known as transcription or basal factors bind to the TATA box and facilitate the binding of RNA polymerase II.

2. **Splice acceptor and donor sequences:** Splice acceptor and donor sequences are one type of consensus sequence found at the 5′ and 3′ ends of introns. Introns nearly always begin with guanine and uracil (GU) nucleotides and end with adenine and guanine (AG) nucleotides, which are preceded by a pyrimidine-rich tract (Figure 8.4). This particular consensus sequence is essential for splicing introns out of the primary transcript.

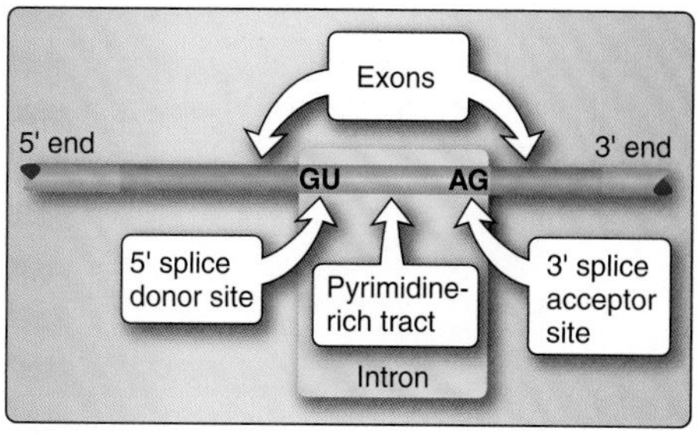

Figure 8.4
Splice acceptor and donor sequences.

IV. RNA SYNTHESIS

Synthesis of RNA from DNA occurs in the nucleus and is catalyzed by an RNA polymerase. RNA differs significantly from DNA in that it is single stranded and contains uracil (U) instead of the thymine (T) found in DNA. Protein-encoding genes produce mRNA as an intermediate to the cytosol for protein synthesis. Regulatory mRNA sequences are important for stability and translational efficiency. These are sequences in the 5′ and the 3′ ends of the mRNA, called untranslated regions (UTR), and are not part of the final protein product.

A. RNA polymerases

There are several distinct RNA polymerases in eukaryotic cells as shown in Table 8.1. The mechanism described below refers to RNA polymerase II, which catalyzes the synthesis of mRNA from protein-coding genes.

B. Several proteins bind to the gene to be transcribed

The reaction catalyzed by RNA polymerase II requires the formation of a large complex of proteins over the start site of the gene. This preinitiation complex is important for accurately positioning the RNA polymerase II on DNA for initiation. This complex consists of general transcription factors and accessory factors.

C. Regulatory regions

An mRNA-producing eukaryotic gene can be divided into its coding and regulatory regions as defined by the transcription start site. The coding region contains the DNA sequence that is transcribed into mRNA, which is ultimately translated into a protein. The regulatory region consists of two classes of sequences (Figure 8.5). One class is responsible for ensuring basal expression and the other for regulated expression.

1. **Basal promoters:** Basal promoter sequences generally have two components. The proximal component, generally the TATA box, directs RNA polymerase II to the correct site and a distal component specifies the frequency of initiation (CAAT and GC boxes).

   The best studied of these is the CAAT box, but several other sequences may be used in various genes. These sequences determine how frequently the transcription event occurs. Mutations in these regions reduce the frequency of transcriptional starts

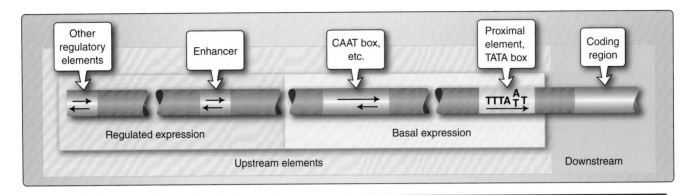

Figure 8.5
Two types of regulatory sequences.

10 to 20 fold. Typical of these DNA sequences are the GC and CAAT boxes, so named because of the DNA sequences involved. These boxes bind specific proteins and the frequency of transcription initiation is a consequence of these protein-DNA interactions, whereas the protein-DNA interaction at the TATA box ensures fidelity of initiation.

2. **Enhancers and response elements:** Enhancers and response elements regulate gene expression. This class consists of sequences that enhance or repress expression and of others that mediate the response to various signals including hormones, chemicals, etc. Depending upon whether they increase or decrease the initiation rate of transcription, they are called enhancers or repressors and have been found both upstream and downstream from the transcription start site. In contrast to proximal and upstream promoter sequences, enhancers and repressors can exert their effects even when located hundreds or even thousands of bases away from the transcription units located on the same chromosome. They also function in an orientation-independent fashion. These regions are bound by proteins (specific transcription factors) that regulate gene expression and are discussed in Chapter 10.

D. Basal transcription complex formation

Basal transcription requires, in addition to RNA polymerase II, a number of transcription factors called A, B, D, E, F, and H, some of which are composed of several different subunits (Figure 8.6). These general transcription factors are conventionally abbreviated as TFII A, B, etc. (transcription factor, class II gene). TFIID (consists of TATA binding protein [TBP] + 8 to 10 TBP-associated factors), which binds to the TATA box, is the only one of these factors capable of binding to specific sequences of DNA. Binding of TBP to the TATA box in the minor groove causes a bend in the DNA helix. This bending is thought to facilitate the interaction of TBP-associated proteins with other components of the transcription initiation complex and, possibly, with other factors bound to the upstream sequences. One of the transcription factors, TFIIF, has DNA helicase activity that promotes the unwinding of the DNA near the transcription start site. This allows the opening of the complex to allow for transcription. RNA polymerase II is also phosphorylated in its C-terminal domain, which allows it to extricate from the promoter and begin elongating a transcript.

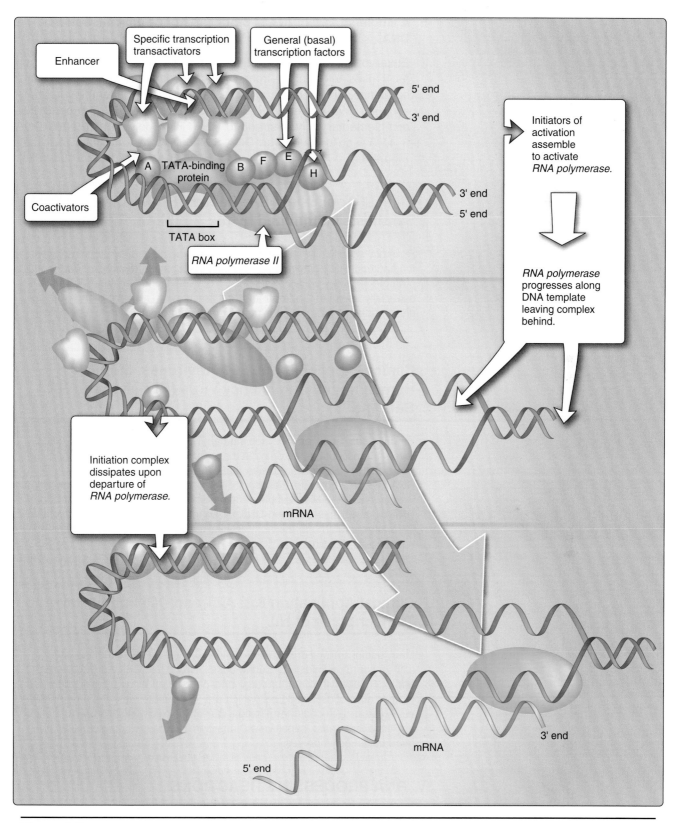

Figure 8.6
Formation of the transcription complex requires several proteins in addition to RNA polymerase II.

E. A single-stranded RNA is produced from a double-stranded DNA

Eukaryotic RNA polymerase is a DNA-dependent RNA polymerase as it uses information from DNA to synthesize a complementary sequence. Only one strand of the gene is used as a template for transcription and is referred to as the template strand. The product is a complementary single-stranded RNA. RNA polymerase reads DNA 3′ to 5′ and produces an RNA molecule complementary to it (see Figure 8.6).

Clinical Application 8.1: Bacterial DNA–Directed RNA Synthesis is Inhibited by the Antibiotic Rifampin

Rifampin specifically inhibits bacterial RNA synthesis by interfering with the bacterial RNA polymerase. The inhibited enzyme remains bound to the promoter, thereby blocking the initiation by uninhibited enzyme. Rifampin is especially useful in the treatment of tuberculosis. This drug along with isoniazid (an antimetabolite) has greatly reduced morbidity due to tuberculosis.

Clinical Application 8.2: Retroviruses, Such as Human Immunodeficiency Virus (HIV), have an RNA Genome

Retroviruses, such as HIV and human T-cell lymphotropic virus, contain reverse transcriptase, an enzyme that copies the RNA genome of the virus into a cDNA. "Reverse" signifies that the biological information flows from RNA to DNA, opposite the usual direction of transfer. Reverse transcriptase mediates the RNA template–dependent information of double-stranded DNA from a single-stranded RNA by an intricate process. The transcribed DNA is integrated into the host cellular genome and is replicated with the host cellular machinery.

Clinical Application 8.3: AZT and DDI Inhibit Reverse Transcriptase of HIV

Many useful antiviral drugs act as antimetabolites because they are structurally similar to pyrimidine or purine bases. Drugs, such as zidovudine (AZT) and ddl (dideoxyinosine), undergo phosphorylation by host cellular kinases to form nucleotide analogues, which are incorporated into the viral nucleic acids resulting in chain termination. Selective toxicity results because viral enzymes are more sensitive to inhibition by these antimetabolites than mammalian polymerases.

V. RNA PROCESSING REACTIONS

Gene transcription produces an RNA that is larger than the mRNA found in the cytoplasm for translation. This larger RNA, called the primary transcript or heterogeneous nuclear RNA (hnRNA), contains segments of transcribed introns. The intron segments are removed and the exons are joined at specific sites, called donor and acceptor sequences, to form the mature mRNA by a mechanism of RNA processing (Figure 8.7).

A. Addition of a 5′ cap

Almost immediately after the initiation of RNA synthesis, the 5′ end of RNA is capped by a methyl guanosine residue, which protects it from degradation (by 5′ exonucleases that digest DNA from a free 5′ or 3′ end) during elongation of the RNA chain. The cap also helps the transcript bind to the ribosome during protein synthesis.

B. Addition of a poly(A) tail

The primary transcripts contain a highly conserved AAUAAA consensus sequence, known as a polyadenylation signal, near their 3′ end. The polyadenylation site is recognized by a specific endonuclease that cleaves the RNA approximately 20 nucleotides downstream. Transcription may proceed for several hundred nucleotides beyond the polyadenylation site, but the 3′ end of the transcript is discarded. The newly created 3′ terminus, however, serves as a primer for enzymatic addition by poly(A) polymerase of up to 250 adenine nucleotides (Figure 8.8). The poly A tail also serves as a mechanism of protection of the mRNA from degradation (Chapter 10).

C. Intron removal

Splice sites are present within the gene and delineate the introns. Splice site sequences, which indicate the beginning (GU) and ending (AG) of each intron, are found within the primary RNA transcript. Introns are removed and exons are spliced (joined) together to form the mature mRNA (Figure 8.9). A special structure called the **spliceosome** converts the primary transcript into mRNA. Spliceosomes comprise the primary transcript, five small nuclear RNAs (U1, U2, U5, and U4/6), and more than 50 proteins. Collectively called snRNPs (pronounced "snurps"), the complex facilitates this process by positioning the RNA for necessary splicing reactions and helps form the structures and intermediates for removal of the intron. The mature RNA molecule now leaves the nucleus by passing into the cytoplasm through the pores in the nuclear membrane.

Clinical Application 8.4: Mutations in Splicing Signals Cause Human Disease

Thalassemias are hereditary anemias that comprise the single most common genetic disorder in the world. The mutations that cause the thalassemia affect the synthesis of either the alpha or the beta chains of globin, causing a decreased production of hemoglobin and, consequently, an anemia. Point mutations can occur within the TATA box or mutations can occur in the splice junction sequences at the intron-exon boundaries.

Some of the splicing abnormalities alter the sequence GT at the beginning of an intron or the AG at the end. Because these sequences are absolutely required for normal splicing, such mutations lead to the loss of beta-globin production. In the case of other mutations that affect the consensus region of the donor or acceptor site, there is a reduced ability of the RNA to correctly splice and it will result in decreased but detectable amounts of beta-globin.

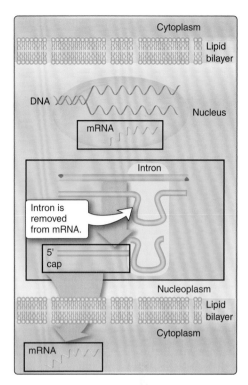

Figure 8.7
mRNA is transcribed and processed in the nucleus.

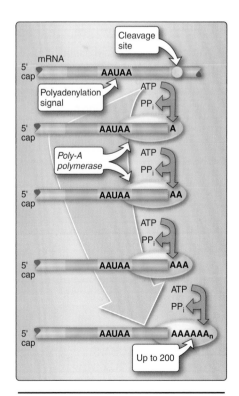

Figure 8.8
RNA processing reactions.

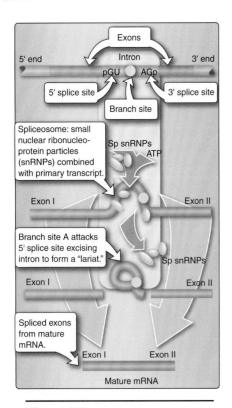

Figure 8.9
mRNA splicing.

Clinical Application 8.5: Transcription-Coupled Repair

TFIIH, a general transcription factor involved in transcription of all genes, also plays a role in nucleotide excision repair in eukaryotic cells. Some of the subunits have homology to helicases, which aid in the unwinding of DNA at the start site during transcription. This sharing of subunits between transcription and repair processes might explain why there is efficient repair of actively transcribed regions more than nontranscribed regions (*transcription-coupled repair*). In this system when there is distortion in DNA, and RNA polymerase is unable to transcribe due to this hindrance, a complex of proteins known as CSA and CSB is recruited to the site. These proteins aid in the opening of the double-stranded DNA and recruitment of the general transcription factor TFIIH, which allow in the opening and subsequent removal of the affected region. CSA and CSB derive their name from **Cockayne syndrome**, a rare inherited disorder where these proteins are defective due to mutations.

Chapter Summary

- RNA polymerase II transcribes protein-coding genes.
- Transcription requires several factors to bind to the regulatory region of the gene.
- Proximal and distal promoters and other regulatory sequences control gene expression.
- RNA is transcribed in the nucleus and undergoes processing before entering the cytoplasm.
- RNA processing reactions include addition of a 5′ methyl guanosine cap, a poly(A) tail, and splicing of introns out of the heterogeneous nuclear transcript.

Study Questions

Choose the ONE best answer.

8.1 What will be the sequence of the single-stranded RNA transcribed from the following segment of double-stranded DNA?

5′-TTGCACCTA-3′
3′-AACGTGGAT-5′

A. 5′-UAGGUGCUU-3′
B. 5′-UUGCACCUA-3′
C. 5′-AACGUGGUA-3′
D. 5′-AUCCACGUU-3′
E. 5′-UUCGUGGAU-3′

Correct answer = B. RNA polymerase reads double-stranded DNA on the template strand (running 3′ to 5′) and synthesizes a complementary single-stranded RNA molecule. RNA contains uracil in place of thymine. So the sequence of the newly synthesized RNA would be similar to the sequence on the coding strand, except in places where thymine occurs.

8.2 Addition of a 5′ 7-methyl guanosine cap to the primary RNA transcript during nuclear processing

   A. Facilitates the assembly of the spliceosome complex.
   B. Identifies the transcript as a transfer RNA molecule.
   C. Protects the RNA against degradation by cellular exonucleases.
   D. Inhibits translation of the RNA molecule into protein.
   E. Prevents RNA molecules from forming double-stranded complexes.

Correct answer = C. A 5′ 7-methyl guanosine group is added to the newly synthesized RNA, and it helps protect the RNA from degradation by enzymes in the cell. The spliceosome complex is assembled around the intron-exon boundary during splicing. Unlike mRNA, tRNA molecules are not modified. The presence of the cap on the mRNA is also important for ribosomes to bind to the mRNA during translation.

8.3 Splicing of a newly synthesized RNA molecule to remove introns and join exons

   A. Occurs in the rough endoplasmic reticulum of the cytosol.
   B. Involves a complex of small nuclear RNA and protein molecules.
   C. Proceeds concurrently with translation.
   D. Is inhibited in bacteria by the drug rifampin.
   E. Is stimulated by the binding of transcription factors to the RNA.

Correct answer = B. Spliceosomes are complexes made up of small nuclear RNA and proteins involved in the process of removal of introns and splicing of exons. The processing reactions take place in the nucleus of the cell, and translation occurs in the cytosol. Rifampin inhibits the initiation of transcription, and bacterial RNA is not processed similar to eukaryotic RNA. The rate of transcription is stimulated by transcription factors.

8.4 Which of the following is an mRNA processing reaction?

   A. Binding of the RNA polymerase to TATA box
   B. Synthesis of an RNA strand using RNA polymerase I
   C. Addition of 7-methyl guanosine residues to the 3′ end of the mRNA
   D. Removal of introns from the heterogeneous nuclear RNA
   E. None of the above

Correct answer = D. RNA processing consists of removal of intron sequences from the newly produced heterogeneous nuclear RNA. Transcription is initiated by the formation of a preinitiation complex. Addition of 7-methyl guanosine occurs in the 5′ end of the mRNA.

8.5 Which of the following sites on a gene is important for recognition of the beginning and the ends of intron sequences?

   A. TATA box
   B. GC box
   C. Splice sites GT and AG
   D. Poly(A) tail
   E. CAAT box

Correct answer = C. GT and AG are sequences recognized at the beginning and the ends of introns and are important during splicing of exons. TATA, GC, and CAAT boxes are promoter sequences, and a poly(A) tail is added to the 3′ end of the mRNA as part of the processing reaction.

Translation

9

I. OVERVIEW

Genetic information that is stored in chromosomes and transmitted to daughter cells through deoxyribonucleic acid (DNA) replication is expressed through transcription to messenger ribonucleic acid (mRNA) and subsequent translation into proteins (Figure 9.1). Protein synthesis is called translation because the "language" of the nucleotide sequence on the mRNA is translated into the language of an amino acid sequence. The process of translation requires a **genetic code**, through which the information contained in the nucleic acid sequence is converted to a specific sequence of amino acids that will fold into a final protein product. Any alteration in the nucleic acid sequence may result in an improper amino acid being inserted into the protein chain, potentially causing disease or even death of the organism. Many proteins are modified after their synthesis, by the covalent addition of phosphates or other groups, causing alteration in activity.

II. THE GENETIC CODE

The genetic code is a dictionary that identifies the correspondence between a sequence of three nucleotide bases or **codons** with a particular amino acid.

A. Codons

Codons are presented in the mRNA language of adenine (A), guanine (G), cytosine (C), and uracil (U). Their nucleotide sequences are always written from the 5′ end to the 3′ end. The four nucleotide bases are used to produce the three-base codons. There are, therefore, 4^3 or 64 different combinations of bases, taken three at a time as shown in Figure 9.2.

1. **How to translate a codon:** This table (or "dictionary") can be used to translate any codon sequence and, thus, to determine which amino acids are coded for by an mRNA sequence. For example, the codon 5′-AUG-3′ codes for methionine (see Figure 9.2). Sixty-one of the 64 codons code for the 20 common amino acids.

2. **Termination ("stop" or "nonsense") codons:** Three of the codons, UAG, UGA, and UAA, do not code for amino acids but rather are termination codons. When one of these codons appears in an mRNA sequence, it signals that the synthesis of the protein coded for by that mRNA is complete.

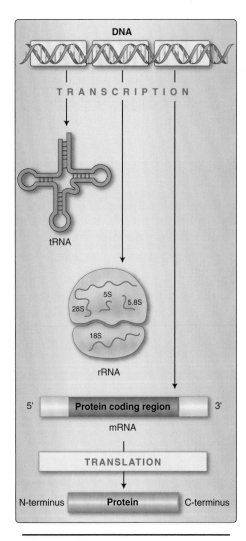

Figure 9.1
Protein synthesis or translation.

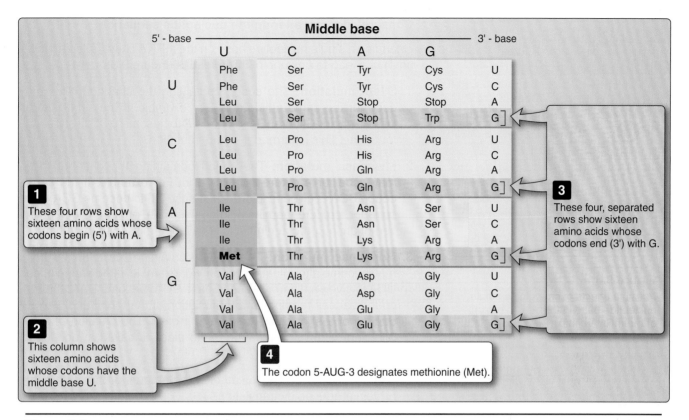

Figure 9.2
Use of genetic code table to translate the codon AUG.

B. Characteristics of the genetic code

Usage of the genetic code is remarkably consistent throughout all living organisms. Characteristics of the genetic code include the following:

1. **Specificity:** The genetic code is specific (unambiguous), that is, a particular codon always codes for the same amino acid.

2. **Universality:** The genetic code is virtually universal, that is, the specificity of the genetic code has been conserved from very early stages of evolution, with only slight differences in the manner in which that code is translated. (Note: An exception occurs in mitochondria, in which a few codons have meanings different than those shown in Figure 9.2, e.g., UGA codes for trp.)

3. **Degeneracy:** The genetic code is degenerate (sometimes called redundant). Although each codon corresponds to a single amino acid, a given amino acid may have more than one triplet coding for it. For example, arginine is specified by six different codons (see Figure 9.2).

4. **Nonoverlapping and commaless:** The genetic code is nonoverlapping and has no punctuation for pauses or commas (commaless). That is, the code is read from a fixed starting point as a continuous sequence of bases, taken three at a time. For example, ABCDEFGHIJKL is read as ABC/DEF/GHI/JKL without any pauses between the codons.

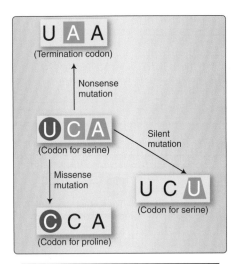

Figure 9.3
Possible effects of changing a single nucleotide base in the coding region of the mRNA chain.

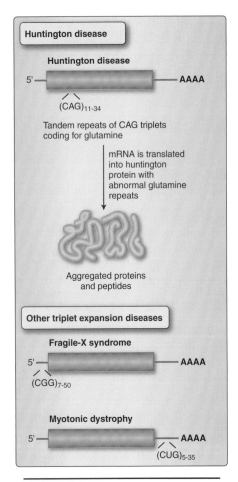

Figure 9.4
Role of tandem triplet repeats in mRNA causing Huntington disease and other triplet expansion diseases.

C. Consequences of altering the nucleotide sequence

Changing a single nucleotide base on the mRNA chain (a "point mutation") can lead to any one of three results (Figure 9.3):

1. **Silent mutation:** The codon containing the changed base may code for the same amino acid. For example, if the serine codon UCA is given a different third base "U" to become UCU, it still codes for serine. This is termed a "silent" mutation.

2. **Missense mutation:** The codon containing the changed base may code for a different amino acid. For example, if the serine codon UCA is given a different first base "C" to become CCA, it will code for a different amino acid, in this case, proline. The substitution of an incorrect amino acid is called a "missense" mutation.

3. **Nonsense mutation:** The codon containing the changed base may become a termination codon. For example, if the serine codon UCA is given a different second base "A" to become UAA, the new codon causes termination of translation at that point and the production of a shortened (truncated) protein. The creation of a termination codon at an inappropriate place is called a "nonsense" mutation.

4. **Other mutations:** These can alter the amount or structure of the protein produced by translation.

   a. **Trinucleotide repeat expansion:** Occasionally, a sequence of three bases that is repeated in tandem will become amplified in number so that too many copies of the triplet occur. If this occurs within the coding region of a gene, the protein will contain many extra copies of one amino acid. For example, amplification of the CAG codon leads to the insertion of many extra glutamine residues in the Huntington protein, causing the neurodegenerative disorder, Huntington disease (Figure 9.4). The additional glutamines result in unstable proteins that cause the accumulation of protein aggregates. If the trinucleotide repeat expansion occurs in the untranslated portion of a gene, the result can be a decrease in the amount of protein produced as seen, for example, in fragile X syndrome and myotonic dystrophy.

   b. **Splice site mutations:** Mutations at splice sites can alter the way in which introns are removed from the pre-mRNA molecules, producing aberrant proteins.

   c. **Frame-shift mutations:** If one or two nucleotides are either deleted from or added to the coding region of a message sequence, a frame-shift mutation occurs and the reading frame is altered. This can result in a product with a radically different amino acid sequence (Figure 9.5) or a truncated product due to the creation of a termination codon. If three nucleotides are added, a new amino acid is added to the peptide, or if three nucleotides are deleted, an amino acid is lost. In these instances, the reading frame is not affected. Loss of three nucleotides maintains the reading frame but can result in serious pathology. For example, cystic fibrosis (CF), a hereditary disease that primarily affects the pulmonary and digestive systems, is most commonly caused by a deletion of three nucleotides from the coding region of a gene, resulting in the loss of phenylalanine at the 508th position (ΔF508) in the protein encoded by that gene. This ΔF508 mutation prevents normal folding of the CF transmembrane conductance regulator (CFTR) protein, leading to its destruction by the proteosome (see Chapter 12). CFTR normally functions as a chloride channel in epithelial cells, and its loss results in the production of thick, sticky secretions in the lungs and pancreas, leading to lung damage and digestive deficiencies. In over 70% of patients with CF, the ΔF508 mutation is the cause of the disease.

III. COMPONENTS REQUIRED FOR TRANSLATION

A large number of components are required for the synthesis of a protein. These include all the amino acids that are found in the finished product, the mRNA to be translated, transfer RNA (tRNA), functional ribosomes, energy sources, and enzymes, as well as protein factors needed for initiation, elongation, and termination of the polypeptide chain.

A. Amino acids

All the amino acids that eventually appear in the finished protein must be present at the time of protein synthesis. (Note: If one amino acid is missing [e.g., if the diet does not contain an essential amino acid], translation stops at the codon specifying that amino acid. This demonstrates the importance of having all the essential amino acids in sufficient quantities in the diet to ensure continued protein synthesis.)

B. Transfer RNA

tRNAs are able to carry a specific amino acid and to recognize the codon for that amino acid. tRNA, therefore, functions as adaptor molecules.

At least one specific type of tRNA is required per amino acid. In humans, there are at least 50 species of tRNA, whereas bacteria contain 30 to 40 species. Because there are only 20 different amino acids commonly carried by tRNA, some amino acids have more than one specific tRNA molecule. This is particularly true of those amino acids that are coded for by several codons.

1. **Amino acid attachment site:** Each tRNA molecule has an attachment site for a specific (cognate) amino acid at its 3′ end (Figure 9.6). The carboxyl group of the amino acid is in an ester linkage with the 3′-hydroxyl group of the ribose moiety of the adenosine nucleotide in the –CCA sequence at the 3′ end of the tRNA. (Note: When a tRNA has a covalently attached amino acid, it is said to be charged; when tRNA is not bound to an amino acid, it is described as being uncharged.) The amino acid that is attached to the tRNA molecule is said to be activated.

2. **Anticodon:** Each tRNA molecule also contains a three-base nucleotide sequence—the anticodon—that recognizes a specific codon on the mRNA (see Figure 9.6). This codon specifies the insertion into the growing peptide chain of the amino acid carried by that tRNA.

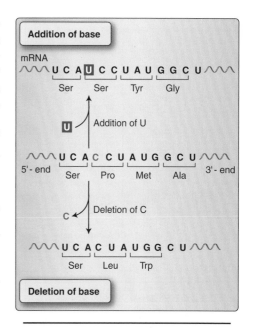

Figure 9.5
Frame-shift mutations as a result of addition or deletion of a base can cause an alteration in the reading frame of mRNA.

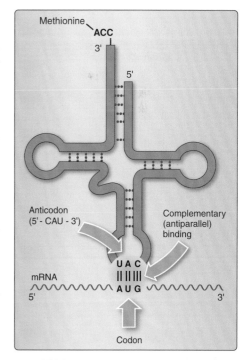

Figure 9.6
Complementary antiparallel binding of the anticodon for methionyl-tRNA (CAU) to the mRNA codon for methionine (AUG).

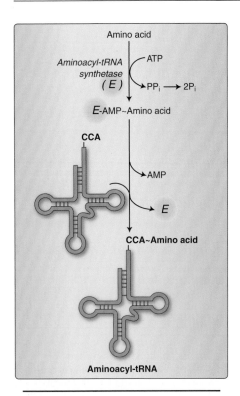

Figure 9.7
Attachment of a specific amino acid to its corresponding tRNA by aminoacyl-tRNA synthetase (E).

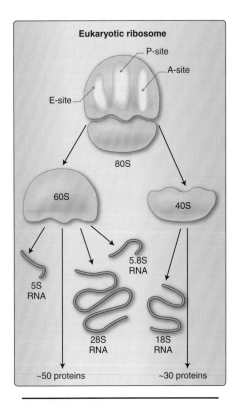

Figure 9.8
Eukaryotic ribosomal composition.

C. Aminoacyl-tRNA synthetases

This family of enzymes is required for the attachment of amino acids to their corresponding tRNA. Each member of this family recognizes a specific amino acid and the tRNA that corresponds to that amino acid (isoaccepting tRNA). These enzymes thus implement the genetic code because they act as molecular dictionaries that can read both the three-letter code of nucleic acids and the 20-letter code of amino acids. Each **aminoacyl-tRNA synthetase** catalyzes a two-step reaction that results in the covalent attachment of the carboxyl group of an amino acid to the 3′ end of its corresponding tRNA. The overall reaction requires adenosine triphosphate (ATP), which is cleaved to adenosine monophosphate (AMP) and inorganic pyrophosphate (PP$_i$) (Figure 9.7). The extreme specificity of the **synthetase** in recognizing both the amino acid and its specific tRNA contributes to the high fidelity of translation of the genetic message. In addition, the synthetases have a "proofreading" or "editing" activity that can remove mischarged amino acids from the enzyme or the tRNA molecule.

D. Messenger RNA

The specific mRNA required as a template for the synthesis of the desired polypeptide chain must be present.

E. Functionally competent ribosomes

Ribosomes are large complexes of protein and ribosomal RNA (rRNA, Figure 9.8). They consist of two subunits—one large and one small—whose relative sizes are generally given in terms of their sedimentation coefficients, or S (Svedberg) values. (Note: Because the S values are determined both by shape as well as molecular mass, their numeric values are not strictly additive. The eukaryotic 60S and 40S subunits form an 80S ribosome.) Prokaryotic and eukaryotic ribosomes are similar in structure and serve the same function, namely, as the "factories" in which the synthesis of proteins occurs.

The larger ribosomal subunit catalyzes the formation of the peptide bonds that link amino acid residues in a protein. The smaller subunit binds mRNA and is responsible for the accuracy of translation by ensuring correct base-pairing between the codon in the mRNA and the anticodon of the tRNA.

1. **Ribosomal RNA:** Eukaryotic ribosomes contain four molecules of rRNA (see Figure 9.8). The rRNAs have extensive regions of secondary structure arising from the base-pairing of the complementary sequences of nucleotides in different portions of the molecule.

2. **Ribosomal proteins:** Ribosomal proteins play a number of roles in the structure and function of the ribosome and its interactions with other components of the translation system.

3. **A, P, and E sites on the ribosome:** The ribosome has three binding sites for tRNA molecules—the A, P, and E sites—each of which extends over both subunits (see Figure 9.8). Together, they cover three neighboring codons. During translation, the A site binds an incoming aminoacyl-tRNA as directed by the codon currently occupying this site. This codon specifies the next amino acid to be added to the growing peptide chain. The P-site codon is occupied

by peptidyl-tRNA. This tRNA carries the chain of amino acids that has already been synthesized. The E site is occupied by the empty tRNA as it is about to exit the ribosome.

4. **Cellular location of ribosomes:** In eukaryotic cells, the ribosomes are either "free" in the cytosol or are in close association with the endoplasmic reticulum (which is then known as the "rough" endoplasmic reticulum, or RER). The RER-associated ribosomes are responsible for synthesizing proteins that are to be exported from the cell as well as those that are destined to become integrated into plasma, endoplasmic reticulum, or Golgi membranes or incorporated in lysosomes. Cytosolic ribosomes synthesize proteins required in the cytosol itself or destined for the nucleus, mitochondria, and peroxisomes. (Note: Mitochondria contain their own set of ribosomes and their own unique, circular DNA.)

F. Protein factors

Initiation, elongation, and termination (or release) factors are required for peptide synthesis. Some of these protein factors perform a catalytic function, whereas others appear to stabilize the synthetic machinery.

G. ATP and GTP are required as sources of energy

Cleavage of four high-energy bonds is required for the addition of one amino acid to the growing polypeptide chain: two from ATP in the **aminoacyl-tRNA synthetase** reaction—one in the removal of PP_i and one in the subsequent hydrolysis of the PP_i to inorganic phosphate by **pyrophosphatase**—and two from guanosine triphosphate (GTP)—one for binding the aminoacyl-tRNA to the A site and one for the translocation step (see Figure 9.10). (Note: Additional ATP and GTP molecules are required for initiation in eukaryotes and an additional GTP molecule is required for termination.)

IV. CODON RECOGNITION BY TRNA

Correct pairing of the codon in the mRNA with the anticodon of the tRNA is essential for accurate translation (see Figure 9.6). Some tRNAs recognize more than one codon for a given amino acid.

A. Antiparallel binding between codon and anticodon

Binding of the tRNA anticodon to the mRNA codon follows the rules of complementary and antiparallel binding, that is, the mRNA codon is "read" 5′ → 3′ by an anticodon pairing in the "flipped" (3′ → 5′) orientation (Figure 9.9). (Note: When writing the sequences of both codons and anticodons, the nucleotide sequence must ALWAYS be listed in the 5′ → 3′ order.)

B. Wobble hypothesis

The mechanism by which tRNAs can recognize more than one codon for a specific amino acid is described by the "wobble" hypothesis in which the base at the 5′ end of the anticodon (the "first" base of the anticodon) is not as spatially defined as the other two bases. Movement of that first base allows nontraditional base-pairing with the 3′ base of the codon (the "last" base of the codon). This movement is called "wobble" and allows a single tRNA to recognize more

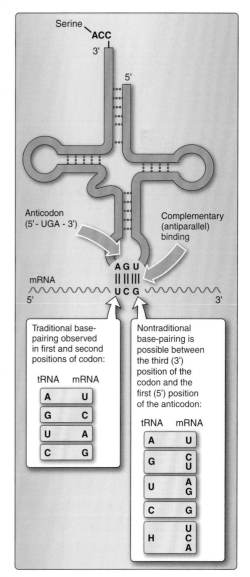

Figure 9.9

Wobble: Nontraditional base-pairing between the 5′-nucleotide (first nucleotide) of the anticodon with the 3′-nucleotide (last nucleotide) of the codon. H, hypoxanthine (the base of inosine).

than one codon. Examples of these flexible pairings are shown in Figure 9.9. The result of wobbling is that there need not be 61 tRNA species to read the 61 codons coding for amino acids.

V. STEPS IN PROTEIN TRANSLATION

The pathway of protein synthesis translates the three-letter alphabet of nucleotide sequences on mRNA into the 20-letter alphabet of amino acids that constitute proteins. The mRNA is translated from its 5′ end to its 3′ end, producing a protein synthesized from its amino-terminal end to its carboxyl-terminal end. The process of translation is divided into three separate steps: initiation, elongation, and termination. The polypeptide chains produced may be modified by posttranslational modification.

A. Initiation

Initiation of protein synthesis involves the assembly of the components of the translation system before peptide bond formation occurs. These components include the two ribosomal subunits, the mRNA to be translated, the aminoacyl-tRNA specified by the first codon in the message, GTP (which provides energy for the process), and initiation factors that facilitate the assembly of this initiation complex (see Figure 9.10). (Note: In prokaryotes, three initiation factors are known [IF-1, IF-2, and IF-3], whereas in eukaryotes, there are over ten [designated eIF to indicate eukaryotic origin]. Eukaryotes also require ATP for initiation.) The mechanism by which the ribosome recognizes the nucleotide sequence that initiates translation is different in eukaryotes and prokaryotes.

In eukaryotes, the initiating AUG is recognized by a special initiator tRNA. Recognition is facilitated by eIFs (eIF2 plus additional eIF). The amino acid–charged initiator tRNA enters the ribosomal P site, and GTP is hydrolyzed to GDP. (Note: The initiator tRNA is the only tRNA recognized by eIF-2, and the only tRNA to go directly to the P site.)

B. Elongation

Elongation of the polypeptide chain involves the addition of amino acids to the carboxyl end of the growing chain. During elongation, the ribosome moves from the 5′ end to the 3′ end of the mRNA that is being translated (see Figure 9.10). Delivery of the aminoacyl-tRNA whose codon appears next on the mRNA template in the ribosomal A site is facilitated by the elongation factors (eEF-1α and eEF-1βγ). These factors function as nucleotide exchange factors, exchanging their GTP for the guanosine diphosphate (GDP) (from GTP hydrolysis). The formation of the peptide bonds is catalyzed by **peptidyltransferase,** an activity intrinsic to the 28S rRNA found in the 60S ribosomal subunit. Because this rRNA catalyzes the reaction, it is referred to as a ribozyme. After the peptide bond has been formed, the ribosome advances three nucleotides toward the 3′ end of the mRNA. This process is known as translocation and requires the participation of eEF-2 and GTP hydrolysis. This causes movement of the uncharged tRNA into the ribosomal E site (before being released) and movement of the peptidyl-tRNA into the P site.

INITIATION

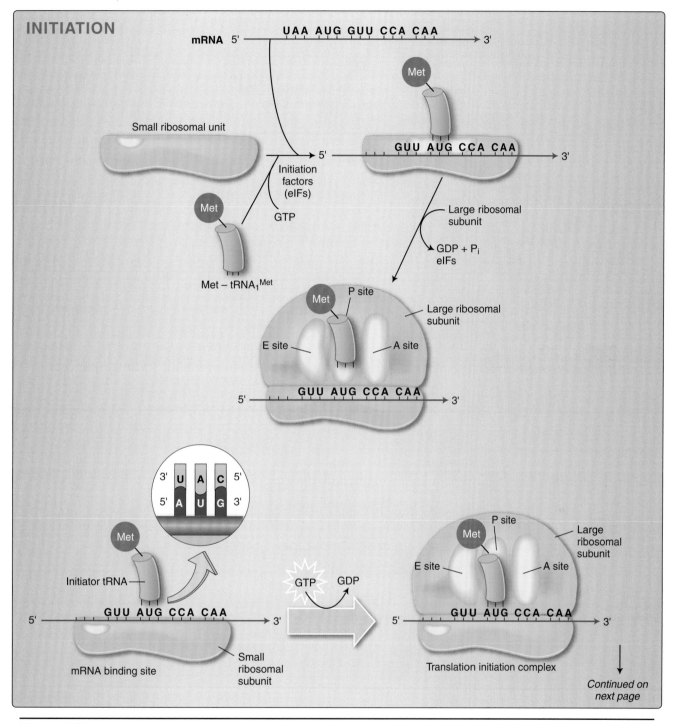

Figure 9.10
Steps in protein synthesis.

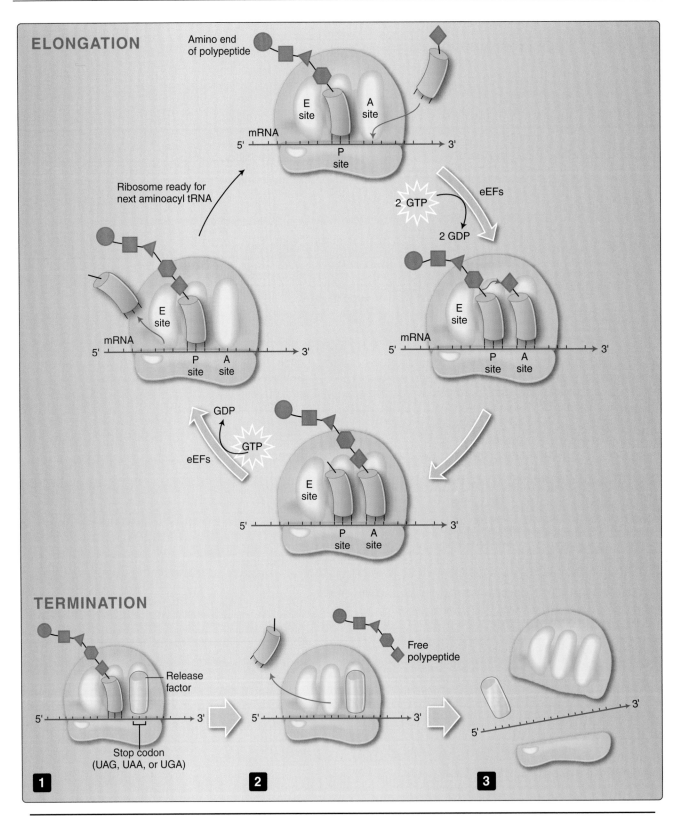

Figure 9.10
Steps in protein synthesis – (continued).

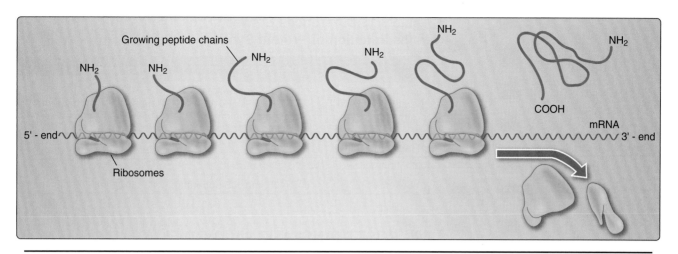

Figure 9.11
A polyribosome consists of several ribosomes simultaneously translating one mRNA.

C. Termination

Termination occurs when one of the three termination codons moves into the A site (see Figure 9.10). Eukaryotes have a single release factor, eRF, which recognizes all three termination codons. The newly synthesized polypeptide may undergo further modification as described below, and the ribosomal subunits, mRNA, tRNA, and protein factors can be recycled and used to synthesize another polypeptide.

D. Polysomes

Translation begins at the 5′ end of the mRNA, with the ribosome proceeding along the RNA molecule. Because of the length of most mRNAs, more than one ribosome at a time can generally translate a message (Figure 9.11). Such a complex of one mRNA and a number of ribosomes is called a polysome or polyribosome.

E. Regulation of translation

Although gene expression is most commonly regulated at the transcriptional level, the rate of protein synthesis is also sometimes regulated. An important mechanism by which this is accomplished in eukaryotes is by covalent modification of eIF-2 (phosphorylated eIF-2 is inactive).

VI. SEVERAL ANTIMICROBIAL ANTIBIOTICS TARGET BACTERIAL TRANSLATION

The initiation process in protein synthesis is different in prokaryotes and eukaryotes as shown in Table 9.1. Many antibiotics that are used to combat bacterial infections in humans take advantage of the differences between the mechanisms for protein synthesis in prokaryotes and eukaryotes (Table 9.2).

Table 9.1: Differences Between Prokaryotes and Eukaryotes in the Initiation of Protein Synthesis

	Eukaryotes	Prokaryotes
Binding of mRNA to small ribosomal subunit	Cap at 5′-end of mRNA binds to eIFs and 40S ribosomal subunit. mRNA is scanned for first AUG	A specific sequence upstream of initiating AUG binds to complementary sequence in 16S RNA
First amino acid	Methionine	Formylmethionine
Initiation factors	eIFs (12 or more)	IFs (3)
Ribosomes	80S (40S and 60S subunits)	70S (30S and 50S subunits)

VII. POSTTRANSLATIONAL MODIFICATION OF POLYPEPTIDE CHAINS

Many polypeptide chains are covalently modified, either while they are still attached to the ribosome or after their synthesis has been completed. Because the modifications occur after the translation is initiated, they are called posttranslational modifications. These modifications may include the removal of a part of the translated sequence or the covalent addition of one or more chemical groups required for protein activity. Some types of posttranslational modifications are listed below.

A. Trimming

Many proteins destined for secretion from the cell are initially made as large, precursor molecules that are not functionally active. Portions of the protein chain must be removed by specialized endoproteases, resulting in the release of an active molecule. The cellular site of the cleavage reaction depends on the protein to be modified. For example, some precursor proteins are cleaved in the endoplasmic reticulum or the Golgi apparatus, others are cleaved in developing secretory vesicles, and still others, such as collagen, are cleaved after secretion. Zymogens are inactive precursors of secreted enzymes (including the proteases required for digestion). They become activated through cleavage when they reach their proper sites of action. For example, the pancreatic zymogen, trypsinogen, becomes activated to **trypsin** in the small intestine.

 The synthesis of enzymes as zymogens protects the cell from being digested by its own products.

Table 9.2: Effects of Antibiotics on Prokaryotic Protein Synthesis

Streptomycin	Inhibits initiation and causes misreading
Tetracycline	Binds to the 30S subunit and inhibits the binding of aminoacyl-tRNAs
Erythromycin	Binds to the 50S subunit and inhibits translocation

B. Covalent alterations

Proteins, both enzymatic and structural, may be activated or inactivated by the covalent attachment of a variety of chemical groups. Examples of these modifications include (Figure 9.12):

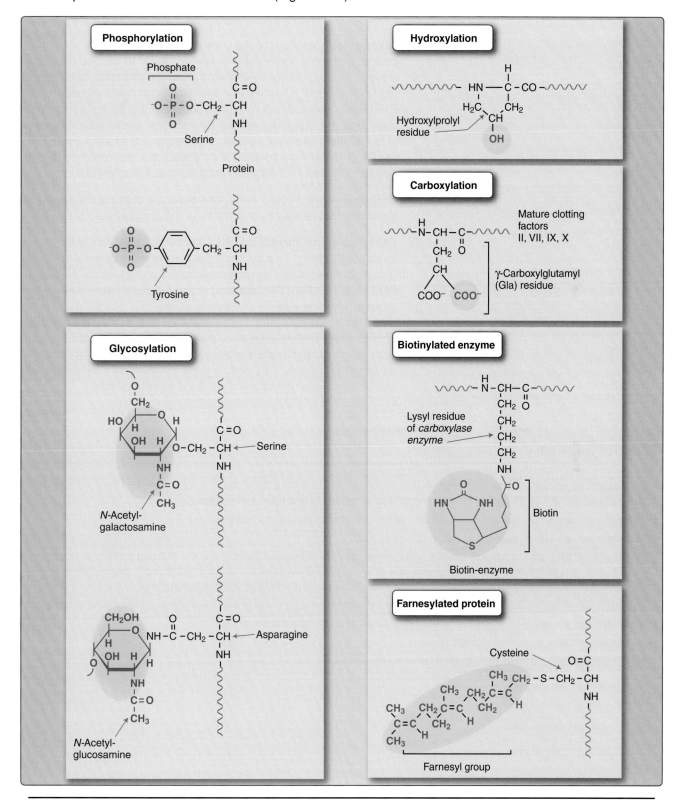

Figure 9.12
Posttranslational modifications of some amino acid residues.

1. **Phosphorylation:** Phosphorylation occurs on the hydroxyl groups of serine, threonine, or, less frequently, tyrosine residues in a protein. This phosphorylation is catalyzed by one of a family of protein kinases and may be reversed by the action of cellular protein phosphatases. The phosphorylation may increase or decrease the functional activity of the protein.

2. **Glycosylation:** Many of the proteins that are destined to become part of a plasma membrane or lysosome, or to be secreted from the cell, have carbohydrate chains attached to serine or threonine hydroxyl groups (O-linked) or the amide nitrogen of asparagine (N-linked). The addition of sugars occurs in the endoplasmic reticulum and the Golgi apparatus. Sometimes, glycosylation is used to target the proteins to specific organelles. For example, enzymes destined to be incorporated into lysosomes are modified by the phosphorylation of mannose residues (see Chapter 11).

3. **Hydroxylation:** Proline and lysine residues of the α chains of collagen are extensively hydroxylated in the endoplasmic reticulum.

4. **Other covalent modifications:** These may be required for the functional activity of a protein. For example, additional carboxyl groups can be added to glutamate residues by vitamin K–dependent carboxylation. The resulting γ-carboxyglutamate residues are essential for the activity of several of the blood-clotting proteins. Biotin is covalently bound to the ε-amino groups of lysine residues of biotin-dependent enzymes that catalyze carboxylation reactions, such as **pyruvate carboxylase**. Attachment of lipids, such as farnesyl groups, can help anchor proteins in membranes. In addition, many proteins are acetylated after translation.

Chapter Summary

- Codons are composed of three nucleotide bases presented in the mRNA language of A, G, C, and U. There are 64 possible combinations with 61 coding for the 20 common amino acids and three termination signals.
- The genetic code is specific, universal, degenerate, nonoverlapping, and commaless.
- Mutations are a result of altering the nucleotide sequence.
- Requirements for protein synthesis include all the amino acids that eventually appear in the finished protein, at least one specific type of tRNA for each amino acid, one aminoacyl-tRNA synthetase for each amino acid, the mRNA coding for the protein to be synthesized, ribosomes, protein factors, and ATP and GTP as energy sources.
- The formation of the peptide bond is catalyzed by peptidyl transferase, an activity intrinsic to the large ribosomal subunit.
- More than one ribosome at a time can translate a message forming a polysome.
- Numerous antibiotics interfere with the process of protein synthesis selectively in prokaryotes and eukaryotes.
- Many polypeptides are covalently modified after synthesis.

Study Questions

Choose the ONE best answer.

9.1 A 20-year-old man diagnosed with anemia is found to have an abnormal form of β-globin that is 172 amino acids long, rather than the 141 found in the normal protein. Which of the following point mutations is consistent with this abnormality?

    A. UAA → CAA

    B. UAA → UAG

    C. CGA → UGA

    D. GAU → GAC

    E. GCA → GAA

Correct answer = A. Mutating the normal stop codon for β-globin from UAA to CAA causes the ribosome to insert a glutamine at that point. Thus, it will continue extending the protein chain until it comes upon the next stop codon further down the message, resulting in an abnormally long protein. A change from UAA to UAG would simply change one stop codon for another and would have no effect on the protein. The replacement of CGA (arginine) with UGA (stop) would cause the protein to be too short. GAU and GAC both encode aspartate and would cause no change in the protein. Changing GCA (alanine) to GAA (glutamate) would not change the size of the protein product.

9.2 A tRNA molecule that is supposed to carry cysteine (tRNA^{cys}) is mischarged, so it actually carries alanine (ala-tRNA^{cys}). What will be the fate of this alanine residue during protein synthesis?

    A. It will be incorporated into a protein in response to an alanine codon.

    B. It will be incorporated into a protein in response to a cysteine codon.

    C. It will remain attached to the tRNA, as it cannot be used for protein synthesis.

    D. It will be incorporated randomly at any codon.

    E. It will be chemically converted to cysteine by cellular enzymes.

Correct answer = B. Once an amino acid is attached to a tRNA molecule, only the anticodon of that tRNA determines the specificity of incorporation. The mischarged alanine will, therefore, be incorporated in the protein at a position determined by a cysteine codon.

9.3 In a patient with cystic fibrosis caused by the ΔF508 mutation, the mutant cystic fibrosis transmembrane conductance regulator (CFTR) protein folds incorrectly. The patient's cells modify this abnormal protein by attaching ubiquitin molecules to it. What is the fate of this modified CFTR protein?

    A. It performs its normal function, as the ubiquitin largely corrects for the effect of the mutation.

    B. It is secreted from the cell.

    C. It is placed into storage vesicles.

    D. It is degraded by the proteosome.

    E. It is repaired by cellular enzymes.

Correct answer = D. Ubiquitination usually marks the old, damaged, or misfolded proteins for destruction by the proteosome. There is no known cellular mechanism for repair of damaged proteins.

9.4 Translation of a synthetic polyribonucleotide containing the repeating sequence CAA in a cell-free protein-synthesizing system produces three homopolypeptides: polyglutamine, polyasparagine, and polythreonine. If the codon for glutamine and asparagine are CAA and AAC, respectively, which of the following triplets is the codon for threonine?

    A. AAC

    B. CAA

    C. CAC

    D. CCA

    E. ACA

Correct answer = E. The synthetic polynucleotide sequence of CAACAACAACAA could be read by the in vitro protein synthesizing system starting at the first C, the first A, or the second A. In the first case, the first triplet codon would be CAA, which codes glutamine; in the second case, the first triplet codon would be AAC, which codes for asparagine; and in the last case, the first triplet codon would be ACA, which codes for threonine.

Regulation of Gene Expression

10

I. OVERVIEW

The deoxyribonucleic acid (DNA) sequence within each and every somatic cell contains all the information required to synthesize thousands of different ribonucleic acid (RNA) molecules and proteins. Typically, a cell expresses only a fraction of its genes as proteins. Different cell types in multicellular organism arise because each type expresses a different set of genes. Moreover, cells can change the pattern of genes they express in response to changes in environment, including signals from other cells. Although all the steps involved in expressing a gene can, in principle, be regulated, for most genes, the initiation of RNA transcription is the most important point of control.

II. STEPWISE REGULATION OF GENE EXPRESSION

There are several potential sites for regulation of gene expression, starting with DNA and transcription to messenger RNA (mRNA) and also including posttranslational modification of newly synthesized protein (Figure 10.1). Epigenetic changes to the genome involve chemical and structural modifications to the chromatin and to DNA, while processing and transport of the newly synthesized mRNA into the cytoplasm are also regulated. In the cytoplasm, stability of the mRNA and its translatability can be controlled. Most proteins are modified after translation controlling their activities, compartmentalization, and half-lives.

A. Transcriptional control

When and how often a gene sequence is copied into RNA is termed transcriptional control, and this occurs at two levels:

- Structural-chemical modifications (i.e., acetylation of histones and demethylation of CpG nucleotides, see Chapter 6) convert compacted chromatin into a less tightly coiled DNA structure (Figure 10.2), allowing access by transcription factors required for gene expression.

- DNA-binding proteins, known as transcription factors, modulate gene expression to turn transcription on or off. There are two categories of transcription factors, general (or basal) and specific.

1. **General (basal) transcription factors:** General transcription factors are abundant proteins that assemble on all genes transcribed by RNA polymerase II (see Chapter 8). These transcription factors

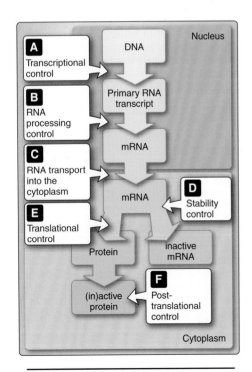

Figure 10.1
Regulation of gene expression can occur at different levels

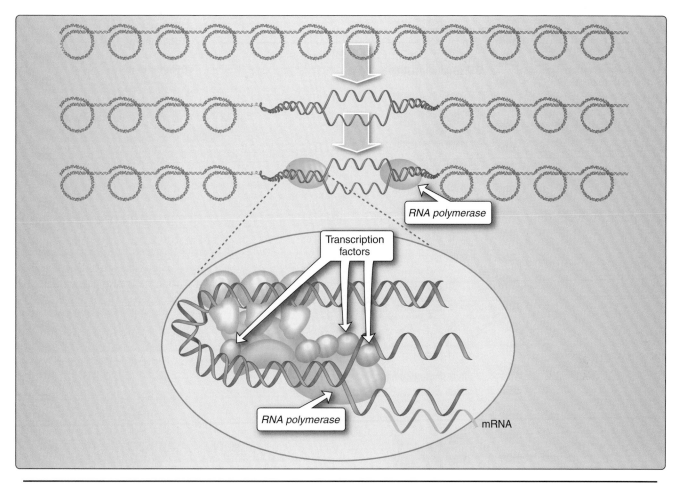

Figure 10.2
Transcription requires DNA to be decondensed.

are important for basal activity of the promoter and to position and activate RNA polymerase II at the start of a protein-coding sequence (Figure 10.3).

2. **Specific transcription factors:** Specific transcriptional factors or gene regulatory proteins are present in very few copies in the individual cells and perform their function by binding to a specific DNA nucleotide sequence and allowing the genes that they control to be activated or repressed. These proteins recognize short stretches of double-stranded DNA of defined sequence and thereby determine which of the thousands of genes in a cell will be transcribed. Many unique regulatory proteins have been identified, each with unique structural motifs, and most bind to the DNA as homodimers or heterodimers (Figure 10.4). The precise amino acid sequence of the motif determines the DNA sequence(s) recognized. Specific transcription factors are important for tissue-specific gene expression and for cell growth and differentiation, and some lipid-soluble hormones regulate the transcription factors in their target cells.

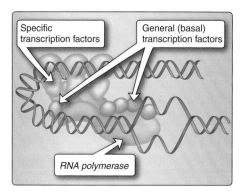

Figure 10.3
General transcription factors are needed for priming transcription.

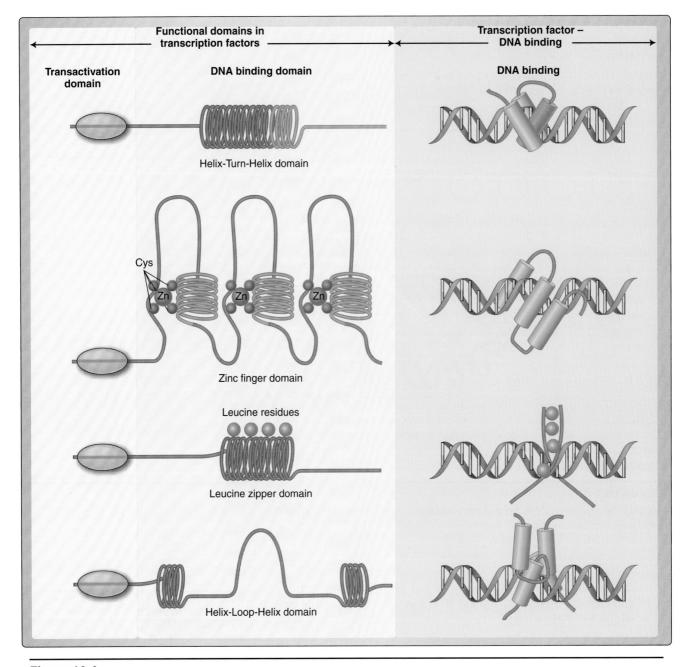

Figure 10.4
Specific transcription factors have a modular design.

Some examples of transcription factors and the sequences recognized by these proteins on DNA are represented in Table 10.1.

RNA polymerase II catalyzes the synthesis of RNA from a DNA template at a singular rate of about 30 to 40 nucleotides per second. Although the rate of synthesis (transcription) is constant, the numbers of polymerases that simultaneously synthesize RNA from a given gene sequence determine the absolute gene transcription rate. Specific transcription factors modulate the number of RNA polymerase molecules actively synthesizing RNA from a

Table 10.1: Specific Transcription Factors are Designed to Bind to Specific DNA Sequences and Regulate Gene Transcription

Transcription Factor	Sequence Recognized
Myc and Max	CACGTG
Fos and Jun	TGACTCA
TR (thyroid hormone receptor)	GTGTCAAAGGTCA
MyoD	CAACTGAC
RAR (retinoic acid receptor)	ACGTCATGACCT

given segment of DNA (Figure 10.5). For this reason, transcription factors have a modular design consisting of at least two distinct domains (DNA-binding and transcription-activating domains). One domain consists of the structural motif that recognizes specific DNA sequences (DNA binding—discussed above) and the other domain contacts the transcriptional machinery and accelerates the rate of transcription initiation by accelerating the assembly of the general transcription factors at the promoter site (transcription activating) (see Figure 10.5).

B. RNA processing control

The primary transcript is produced as heterogeneous nuclear RNA containing introns, which are eventually spliced out to create the mature mRNA (see Chapter 8). This process occurs in the nucleus, and the subsequent processing is necessary to control the number of mRNA molecules that are eventually translated.

1. **mRNA capping:** The addition of the 5′ cap structure (see Chapter 8) is critical for an mRNA to be translated in the cytoplasm and is also needed to protect the growing RNA chain from degradation in the nucleus by 5′ exonucleases.

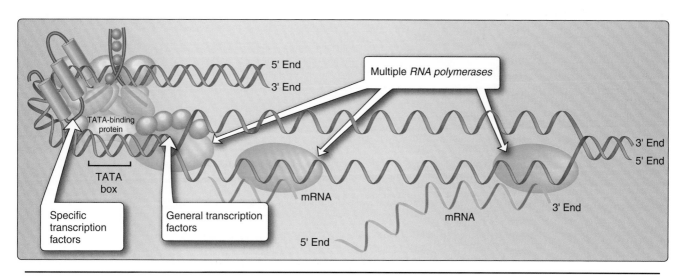

Figure 10.5
Specific transcription factors influence the number of RNA polymerases that bind to DNA and initiate transcription.

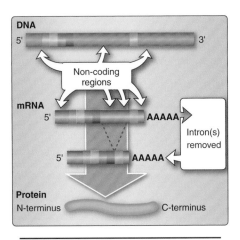

Figure 10.6
RNA processing reactions.

2. **Poly(A) tail:** The second modification of an mRNA transcript occurs at its 3′ end, the addition of a poly(A) tail (approximately 200 adenine nucleotide residues are added). The polyadenylation reaction is an important regulatory step because the length of poly(A) tail modulates both mRNA stability and translation efficiency. The poly(A) tail protects the mRNA from premature degradation by 3′ exonucleases.

3. **Removal of introns:** Following the modification of the 5′ and 3′ ends of the primary transcript, the noninformational intron segments are removed and the coding exon sequences joined together by RNA splicing (Figure 10.6). The specificity of exon joining is conferred by the presence of signal sequences marking the beginning (5′ donor site) and the end (3′ acceptor site) of the intron segment. As these signal sequences are highly conserved, alterations in these sequences can lead to aberrant mRNA molecules.

4. **Alternative splicing:** The ability of the genes to form multiple proteins by joining different exon segments in the primary transcript is called alternative splicing. Alternative splicing is made possible by changing the accessibility of the different splice sites to the splicing machinery by RNA-binding proteins. These proteins could mask preferred splice sites or change local RNA structure to promote the splicing of alternate sites. In addition, cell-specific regulation can determine the type of alternate transcript and eventually the protein product produced. RNA splicing also allows switching between the production of nonfunctional and functional proteins, membrane-bound protein versus secreted protein, etc. The ability to make more than one protein product from a gene may also explain why the human genome has fewer genes than expected (Figure 10.7).

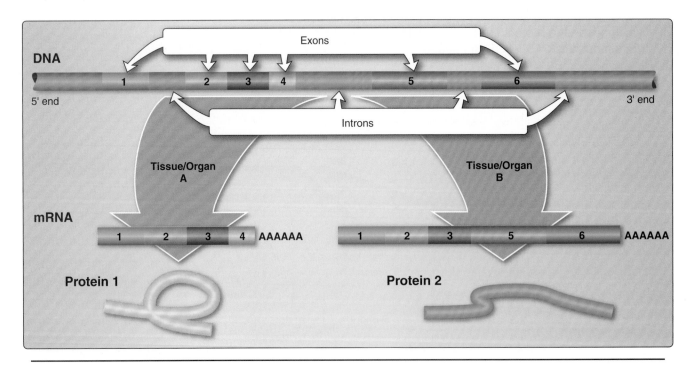

Figure 10.7
Alternative splicing of genes to generate multiple proteins.

Clinical Application 10.1: Alternative Splicing of the Calcitonin Gene

Alternative splicing produces two different proteins from the calcitonin gene. In the parafollicular cells of the thyroid gland, the calcitonin gene produces an mRNA that codes for the calcium-regulating hormone calcitonin, which counteracts the action of parathyroid hormone. Calcitonin is used in the treatment of postmenopausal osteoporosis in women when estrogen is contraindicated. In neural tissues, the same calcitonin gene is spliced differently and uses a different polyadenylation site to give rise to a neuropeptide, calcitonin gene–related peptide (CGRP), which plays a central role in the pathophysiology of migraine. The serum levels of CGRP are elevated in patients during all forms of vascular headaches, including migraines and cluster headaches. These findings suggest that CGRP-receptor antagonists might be effective in treating migraine by blocking CGRP activity.

C. RNA transport into the cytoplasm

The fully processed mRNAs represent only a small proportion of the RNA found in the nucleus. Damaged and misprocessed RNAs are retained within the nucleus and degraded. A typical mature mRNA carries a collection of proteins that identifies it as being the mRNA destined for transport. Export takes place through nuclear pore complex, but mRNAs and their associated proteins are large and require active transport. Different kinds of RNA use different nuclear export pathways. In the case of non–protein-coding RNAs, a class of protein transport receptors known as karyopherins mediate the movement of these RNAs and require the aid of a small GTP-hydrolyzing protein called Ran. Spliced mRNA crosses through the nuclear pore via a Ran-independent pathway that requires a distinct set of export factors. These proteins associate to form a large complex called TREX (transcription-export). The TREX complex interacts with RNA polymerase II and facilitates the loading of associated factors on the nascent RNA for packaging and nuclear export, thus integrating all the steps in mRNA biogenesis. Some of the proteins in the TREX complex can interact with the proteins associated with the 5′ cap of the mRNA thus transporting them out of the nucleus in the 5′-3′ direction (Figure 10.8).

D. Stability control

The length of time mRNAs remain in the cytosol determines the amount of protein product produced by translation. All transcripts have a finite lifetime in the cell. The steady-state level of individual RNA species in a cell is determined by both the rate of transcription and the rate of decay.

1. **mRNA half-life:** In eukaryotic cells, mRNAs are degraded at different selective rates. RNA degradation is carried out by ribonucleases that hydrolyze RNAs to their component nucleotides. A measure of degradative rate for a particular mRNA is called the

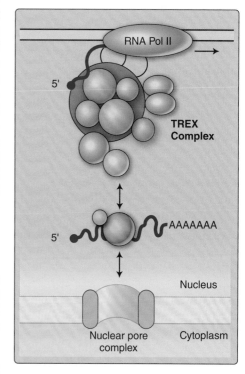

Figure 10.8
mRNA export from nucleus to the cytoplasm.

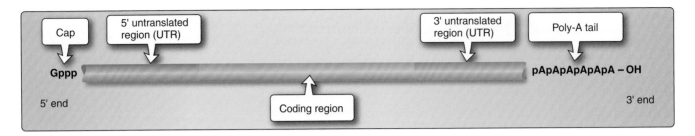

Figure 10.9
Untranslated regions on mRNA.

half-life (the period it takes to degrade an RNA population to half its initial concentration). The unstable mRNAs usually code for regulatory proteins whose production levels change rapidly within cells (e.g., growth factors and gene regulatory proteins) (half-life—minutes to hours). The stable mRNAs usually code for housekeeping proteins (half-life in days).

2. **mRNA 3′UTRs:** The mRNA contains regions that are not translated, and these include the 5′ untranslated region (5′UTR) and the 3′ untranslated region (3′UTR). The stability of an mRNA can be influenced by signals inherent in an RNA molecule. The sequence AUUUA, when found in the 3′UTR, is a signal for early degradation (and therefore short lifetime). The more times the sequence is present, the shorter the lifespan of the mRNA. Because it is encoded in the nucleotide sequence, this is a set property of each different mRNA. Sequences in the 3′UTR form a stem-loop structure that allows for protein binding and protection from degradation (Figure 10.9).

Clinical Application 10.2: Regulation of Iron Levels

Iron deficiency continues to be one of the most prevalent nutritional deficiencies throughout the world. The diagnosis of iron deficiency is based primarily on laboratory measurements. Since free iron is reactive and therefore toxic to the body, most of the iron exists within the body bound to proteins. Free iron is bound to storage proteins (ferritin) or is transported bound to transferrin. Although ferritin is mostly confined to the intracellular compartment, trace amounts are secreted into the blood. Because plasma ferritin is proportional to that stored intracellularly, consequently, plasma ferritin is a valuable index of intracellular stores in iron deficiency anemia. Transferrin receptor levels are also measured clinically as the rate of internalization of iron by cells is determined by its level of cell surface expression. Ferritin and transferrin receptor biosynthesis are inversely modified based on cellular iron levels. Transferrin receptor biosynthesis is increased when available iron levels are low and decreased when iron levels are high. As shown in Figure 10.10, this regulation is accomplished by altering the amount of transferrin receptor and ferritin mRNA. The iron response elements on the transferrin receptor and ferritin mRNA are bound by the iron response protein whose activity is dependent on the iron status of the cell.

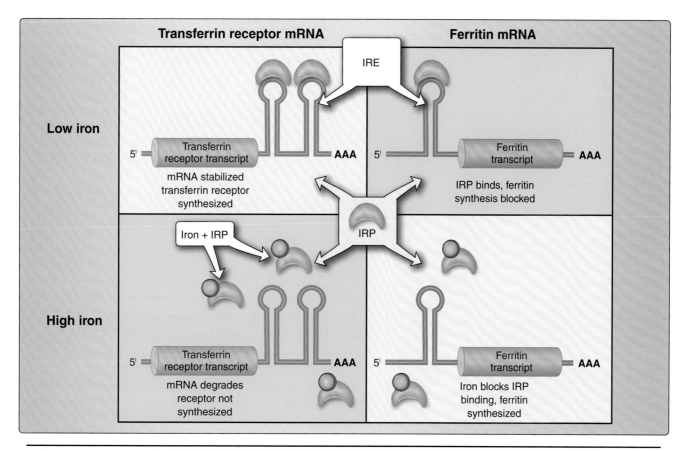

Figure 10.10
Regulation of transferrin receptor and ferritin mRNA.

E. Translational control

The second basic step in gene expression is the translation of mRNAs into protein. Translation occurs in the cytoplasm; however, not all mRNAs are translated on arrival. The RNA molecules in the cytoplasm are constantly associated with proteins, some of which may function to regulate translation. Mechanisms for translational repression are most widely operative. In the case of ferritin, its mRNA is maintained within the cytoplasm but inhibited from being translated until intracellular concentration of iron rises. The block in this case is mediated by the binding of a repressor protein (the same protein that binds transferrin receptor mRNA in the 3′UTR) to the 5′ nontranslated end of the molecule (see Figure 10.10).

F. Posttranslational control

Once cytoplasmic ribosomal complexes have synthesized a protein, the functional capabilities of the protein are often not realized until the protein has been modified. Although these mechanisms, collectively known as posttranslational modifications, are not thought of as gene expression controls, it is recognized that the manifestation of gene expression is not complete until a protein is carrying out its function within the cell.

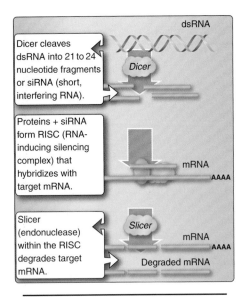

Figure 10.11
RNA interference or RNAi.

III. RNA INTERFERENCE

The presence of double-stranded (ds) RNA in a eukaryotic cell can trigger a process known as RNA interference or RNAi (also known as RNA silencing or RNA inactivation). RNAi consists of two main phases. First, the double-stranded RNA (dsRNA) is recognized by an endonuclease (*Dicer*) and cleaved into smaller molecules of 21 to 24 nucleotides called short interfering RNA (siRNA). In the second phase, a single strand (the guide or antisense strand) of the siRNA associates with proteins to form an RNA-induced silencing complex, or RISC. The guide strand component of RISC then hybridizes with a complementary sequence of a full-length target mRNA. An endonuclease (*Slicer*) in the RISC degrades the target mRNA (Figure 10.11). RNAi is thought to be a part of the body's natural immune system evolved as a defense against retroviruses, such as the human immunodeficiency virus (HIV), that store their genetic information in dsRNA.

Clinical Application 10.3: Antisense RNA Oligonucleotides and Inhibitory RNA (RNAi) as Potential Chemotherapeutic Agents

Antisense RNA is a single-stranded RNA that is complementary to an mRNA designed to inhibit specific mRNA translation by its ability to base pair with the mRNA and physically obstruct the translation machinery from accessing the mRNA. This approach was developed in the 1990s to overcome the limitations of cytotoxic chemotherapy using DNA intercalating drugs and antimetabolites, which do not discriminate between normal and cancer cells. Historically, this approach has not yielded results due to the effects of RNA interference, a process that has more recently been elucidated. But this technology has been effective in the creation of the first antisense RNA drug Vitravene (fomivirsen) used to treat cytomegalovirus-induced retinitis in immune-compromised AIDS patients. The drug is a periodically administered intravitreal injection and is claimed to cause only mild side effects as compared to some other antiviral drugs. RNA interference of exogenously supplied double-stranded RNA offers substantial therapeutic potential. Applications that are in clinical trials include siRNA treatment for age-related macular degeneration (AMD), siRNAs against the respiratory syncytial virus, and a phase I trial to treat patients infected with HIV.

Chapter Summary

- Eukaryotic gene expression can be regulated at different levels of gene expression, including transcription, processing, mRNA stability, and translation.

- Transcription can be controlled by several transcription factors and is an important level of control. While general transcription factors are required for basal expression, specific transcription factors increase the transcription of a gene over basal levels when bound to enhancers and other response elements. They are also important for tissue-specific gene expression.

- Some RNA molecules can undergo differential splicing, producing different mRNA molecules encoding slightly different polypeptides.

Chapter Summary (Continued)

- Transcription factors have one functional domain for DNA binding and one for transcription activation. Transcription factors can be classified according to the structure of their DNA-binding domains; these include zinc finger proteins, helix-turn-helix proteins, leucine zipper proteins, helix-loop-helix proteins, and steroid receptors. Transcription factors affect the number of RNA polymerases binding DNA.

  o In the cytoplasm, translation can be controlled by both the ability of ribosomes to bind mRNA and by the stability of the mRNA.

  o Sequences in the 3′UTR control stability and sequences in the 5′UTR control translation efficiency.

- Proteins are also made active by posttranslational modifications that include phosphorylation, gamma carboxylation, etc.

Study Questions

Choose the ONE best answer.

10.1 If a newly discovered protein is found to contain leucine zipper domains, the putative function of this protein is in

  A. Binding the specific DNA sequences.
  B. Cleaving the mRNA in the nucleus.
  C. Posttranslational modification of the newly synthesized proteins.
  D. Regulating the poly(A) tail of the mRNA.
  E. Regulating the mRNA's half-life.

Correct answer = A. Leucine zipper domains represent amino acid arrangement in transcription factor proteins that bind DNA. It is a motif that is present in the DNA-binding region of a class of transcription factors. Transcription factors do not behave as exonucleases or endonucleases that cleave nucleic acids. They do not directly influence posttranslational changes in newly synthesized proteins or regulate other aspects of the mRNA structure, such as the poly(A) tail or half-life.

10.2 If you replace the 3′UTR sequence of an mRNA with a half-life of 20 min with the 3′UTR of an mRNA with a half-life of 10 h, then the resultant mRNA will have a half-life of

  A. 10 min.
  B. 20 min.
  C. 5 h and 10 min.
  D. 10 h.
  E. 10 h and 20 min.

Correct answer = D. As signals for the mRNA half-life are inherent to the 3′UTR of the mRNA, switching this region between mRNAs with differing half-lives will produce an mRNA with the corresponding half-life. It will not be a product, nor will it change the lifespan of the mRNA to something other than what the sequence encoded by the 3′UTR specifies.

10.3 Tamoxifen, a drug used to treat breast cancer, is a competitive inhibitor of estrogen receptor, a zinc finger protein. Tamoxifen will therefore affect gene expression in these cells mainly by

  A. Binding to TATA boxes of estrogen-responsive genes.
  B. Changing the splice sites within the estrogen-responsive genes.
  C. Exporting estrogen-responsive mRNAs into the cytoplasm.
  D. Preventing the transcription of estrogen-responsive genes.
  E. Rapidly degrading the estrogen-regulated proteins.

Correct answer = D. Since tamoxifen is an inhibitor of the estrogen receptor, a zinc finger protein, and a specific transcription factor, it will affect the gene at the transcriptional level. General transcription factors bind around the TATA box of the gene to prime transcription. An inhibitor of transcription may not affect transport or lifespan of target proteins.

10.4 In iron deficiency anemia, ferritin levels are low because ferritin mRNA is

    A. Degraded rapidly in the cytoplasm.
    B. Not transcribed from the ferritin gene.
    C. Prevented from being translated.
    D. Retained in the nucleus.
    E. Transcribed only in small amounts.

Correct answer = C. Since there is a need to respond quickly to the changes in iron levels, the ferritin mRNA is always made but is blocked from being translated due to the binding of proteins in the 5′UTR. Ferritin mRNA is always present in the cytoplasm but not degraded. For reasons mentioned above, the mRNA is transcribed from the ferritin gene. If the regulation is at the level of transport, or RNA processing, the mRNA may be retained in the nucleus. Regulation at the level of transcription is not the primary mode for this mRNA.

10.5 Therapeutic antisense RNA oligonucleotides, such as fomivirsen, bind to a complementary sequence in

    A. Genomic DNA and prevent its transcription into RNA.
    B. Messenger RNA and prevent its translation into protein.
    C. Small nuclear RNA and prevent assembly of the spliceosome complex.
    D. Ribosomal RNA and prevent assembly of ribosomes.
    E. Heterogeneous nuclear RNA and prevent its polyadenylation.

Correct answer = B. Antisense RNA oligonucleotides are designed to bind to the target single-stranded mRNA and block translation. They cannot bind to the double-stranded genomic DNA. They will not affect splicing of the mRNA or its processing. They do not have a direct effect on ribosome assembly.

Protein Trafficking

11

I. OVERVIEW

Proteins are synthesized on either free ribosomes or ribosomes bound to endoplasmic reticulum (ER) (see also Chapter 5 for a discussion of organelles). Ribosomes are directed to bind to the ER when they are involved in synthesizing proteins destined for insertion into cell membranes, for function within lysosomes, and for secretion outside the cell (Figure 11.1). The newly synthesized proteins will be modified within the ER and trafficked or moved through a membrane-enclosed transport vesicle to the Golgi complex where additional modifications will be made. Structural features within the protein being produced are recognized by the organelles, facilitating the movement of the protein. These structural features are called **signal sequences** and direct the protein to locations where it can be modified properly in order to become functional. The sequences act as "address labels" routing the new proteins to their proper destinations (see also *LIR Biochemistry*, pp. 166–169). Proteins that will function in the nucleus, mitochondria, or peroxisomes are synthesized on free ribosomes (Figure 11.2). These proteins also contain structural features that enable them to be taken into the organelle where they will function.

In all cases, if the appropriate signals are not incorporated within the new proteins, then the proteins will follow a **default pathway**. Proteins synthesized on bound ribosomes will be secreted from the cell unless they contain the proper signal to direct them to an intracellular location. For proteins synthesized on free ribosomes, the default is to remain in the cytosol.

II. TRAFFICKING OF PROTEINS SYNTHESIZED ON BOUND RIBOSOMES

The presence of a particular amino acid sequence within a newly synthesized protein causes the ribosome synthesizing it to bind to the ER. This sequence is an **N-terminal** (at the amino-terminal region of the protein) **hydrophobic** (containing amino acids that do not interact with water) **signal sequence**, sometimes called a leader sequence. When the leader sequence is present in a newly synthesized protein (nascent polypeptide) still attached to its ribosome, cytosolic compounds composed of protein and RNA known as **signal recognition particles** (SRPs) facilitate the ribosome's attachment to the ER (Figure 11.3). SRPs and the signal sequence together bind to an SRP receptor on the membrane of the ER (see also Chapter 5). The ribosome then docks with the ER membrane, and the new protein enters into the space or lumen between membranes of the ER.

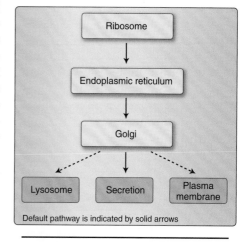

Figure 11.1
Trafficking of proteins synthesized on bound ribosomes.

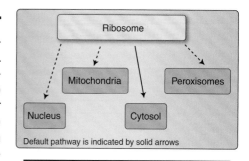

Figure 11.2
Trafficking of proteins synthesized on free ribosomes.

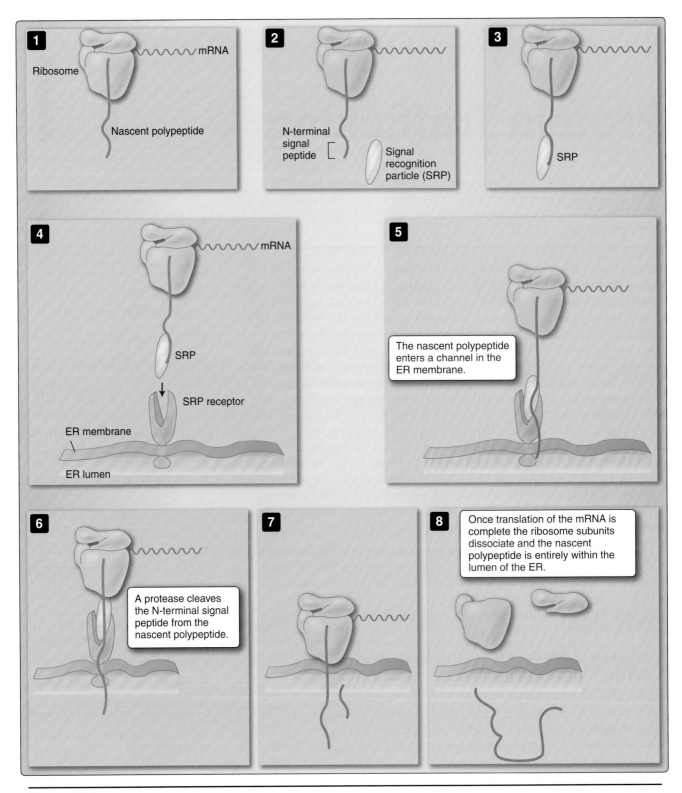

Figure 11.3
Attachment of ribosomes to ER.

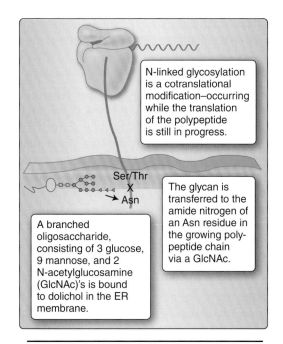

Figure 11.4
Glycosylation in the lumen of the ER.

A. Endoplasmic reticulum

Once a ribosome binds the ER membrane and the nascent polypeptide is translocated into the ER lumen, the signal sequence is removed from the protein by the action of proteases. The remainder of the amino acid sequence of the new protein enters into the lumen, while its ribosome is still bound and synthesizing additional components of the protein. Modifications made to a new protein while it is being translated, such as N-linked glycosylations, are termed **cotranslational processes**.

Addition of carbohydrate or glycosylation occurs to most new proteins that enter into the lumen of the ER. If a nascent polypeptide contains either of two or three amino acid consensus sequences, then carbohydrate will be transferred to an amino acid residue within the new protein. The sequences are Asn-X-Ser and Asn-X-Thr, where Asn is asparagine, X is any amino acid except Proline, Ser is serine, and Thr is threonine. A branched oligosaccharide (often called glycan) is transferred from the membrane lipid, **dolichol**, to the amide nitrogen of Asn in a process known as **N-linked glycosylation** (Figure 11.4). The core glycan consists of 14 residues: 3 glucose, 9 mannose, and 2 *N*-acetylglucosamine (**GlcNAc**), with GlcNAc attaching to the Asn.

Once translation is complete, the ribosome dissociates from the ER. While some new proteins within the ER lumen are destined to remain and function in the ER, most will continue to traffic onward to the Golgi complex. Such proteins make their way next through the series of membrane spaces within the ER to an area of smooth ER (lacks attached ribosomes) known as the **transitional element**. This region of the ER facilitates transfer of the nascent polypeptide to the next organelle in its path, the Golgi complex. The membrane of the transitional element surrounds and encloses the nascent polypeptide until it buds off from the ER to become a **transport vesicle** (Figure 11.5).

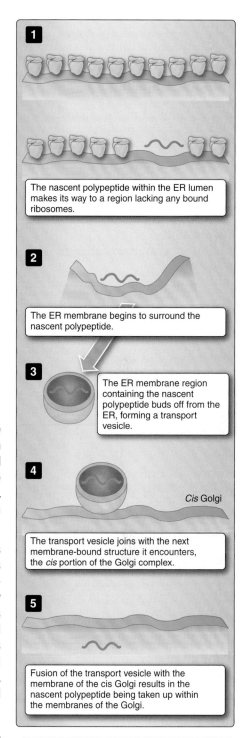

Figure 11.5
Trafficking from ER to Golgi complex in transport vesicles.

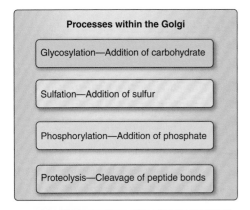

Processes within the Golgi

Glycosylation—Addition of carbohydrate

Sulfation—Addition of sulfur

Phosphorylation—Addition of phosphate

Proteolysis—Cleavage of peptide bonds

Figure 11.6
Modifications of proteins within the Golgi complex.

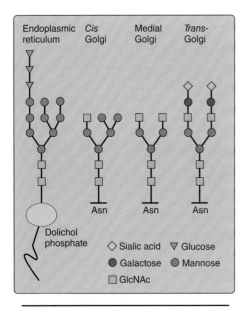

Figure 11.7
Processing in the Golgi.

The vesicle next fuses with the cis Golgi and deposits the nascent polypeptide within the confines of that first region of the Golgi complex.

B. Golgi complex

The Golgi complex is a series of flat, stacked, membranous sacs with three main regions, **cis, medial,** and **trans** Golgi. The nascent polypeptide that enters the cis Golgi will be transferred to the medial and later to the trans Golgi via transport vesicles. Each region is responsible for performing distinct modifications including glycosylation, phosphorylation, sulfation, and proteolysis (enzyme-mediated breakdown of protein) to the proteins being processed (Figure 11.6). For example, **O-linked glycosylation** occurs in the Golgi when carbohydrates are attached to hydroxyl groups of serine or threonine amino acids within Asn-X-Ser/Thr sequences of the nascent polypeptide.

The 14-sugar glycan attached to proteins that received N-linked glycosylation in the ER is processed and altered in the Golgi. In the cis, medial, and then trans Golgi (Figure 11.7), the glycan is first trimmed of all its glucose and several of its mannose residues prior to addition of new GlcNAc residues, then galactose, and finally sialic acid residues. The final glycan attached to proteins that reach the trans Golgi network contains 4 GlcNAc, 3 mannose, 2 galactose, and 2 sialic acid residues.

Some proteins will remain in the Golgi and function in the processing of other new proteins that pass through the Golgi. But most will be modified and sent onward. Many proteins destined to function within lysosomes are phosphorylated on attached mannose, through action of *N*-acetylglucosamine-1-phosphotransferase (GlcNAc-1PT), creating **mannose-6-phosphate** (M6P) tags on proteins destined to function as degradative enzymes within lysosomes (Figure 11.8A).

C. Trans Golgi network and onward

The **trans Golgi network** (TGN) is the final sorting and packaging region of the Golgi. From there, nascent polypeptides are sent either to a lysosome or to the outside of the cell.

1. **Lysosomes:** Lysosomes are membrane-enclosed organelles with acidic internal pH that contain potent degradative enzymes known collectively as **acid hydrolases** (see also Chapter 5). These enzymes function within the acidic environment of lysosomes to hydrolyze nonfunctional macromolecules (proteins, nucleic acids, carbohydrates, and lipids). The majority of acid hydrolase precursors are modified with M6P tags earlier within the Golgi complex. In order to segregate these proteins from other cellular proteins in the Golgi and to ensure that they will be incorporated within a lysosome, receptors for M6P are located in certain regions of the TGN bound by the coat protein, **clathrin** (Figure 11.8B). M6P-containing proteins bind to these receptors and then the TGN that contains the new acid hydrolases bound to M6P receptors buds off, enclosing the new lysosomal proteins in a transport vesicle. Vesicles bound for lysosomes fuse with **endosomes**, which are transport vesicles from the plasma membrane created by endocytosis. The pH is reduced within the precursor lysosome by the

pumping in of protons (H⁺). Next, the clathrin coat is lost and the new proteins dissociate from their M6P receptors, which are recycled back to the TGN for later use. Phosphate is removed from the mannose of the M6P bound to acid hydrolase precursors, allowing the acid hydrolases to become functional enzymes within the lysosome.

Since the default pathway for proteins synthesized on bound ribosomes is to be secreted from the cell, defects in tagging of precursor acid hydrolases will result in their secretion from the cell as nonfunctional proteins. For example, loss of GlcNAc-1PT activity results in secretion of acid hydrolases instead of their incorporation into lysosomes.

Clinical Application 11.1: Mannose-6-Phosphate–Independent Lysosomal Trafficking of Glucocerebrosidase

In the absence of GlcNAc-1PT activity, when acid hydrolase precursors are expected to be secreted constitutively instead of reaching lysosomes, the lysosomal acid hydrolase β-glucosidase can still reach lysosomes, suggesting a signal other than M6P as its tag for lysosomal trafficking. Instead, of using the M6P tagging system, β-glucosidase precursors bind to lysosomal integral membrane protein type 2 (LIMP-2) in a pH dependent manner. Recent studies have confirmed that LIMP-2 is not a substrate for GlcNAc-IP, and not M6P is contained within LIMP-2. Therefore, β-glucosidase, the enzyme deficient in individuals with Gaucher disease, traffics to lysosomes completely independently of M6P and of M6P receptors.

Inherited as an autosomal recessive disorder, Gaucher disease is caused by mutations in the β-glucosidase gene (*GBA*) that normally encodes β-glucocerebrosidase. This disorder is characterized by accumulation of glucocerebroside in macrophages and their collection in the spleen, liver, bone marrow, kidneys, lung, and/or brain, depending on the type of disorder. Considered the most common of the lysosomal storage diseases, Gaucher disease is found in three different forms, types I, II, and III, depending on the nature of the inherited mutations. Type I is the most common and affected individuals show skeletal weakness and have sphingolipid-engorged bone marrow cells that cause deficiencies in both red and white blood cells. Types II and III have neurological involvement, with type II usually resulting in death by age 2, while those with type 1 and type III live into adulthood.

---

2. **Secretion from the cell:** New proteins that leave the TGN and are not destined to function in lysosomes or to be inserted into plasma membranes will be secreted from the cell. Many proteins are released from the cell as soon as their transport vesicle can fuse with the plasma membrane. Other proteins are stored within the cytoplasm within their transport vesicle (sometimes called granules) until the appropriate time for their release from the cell.

   a. **Constitutive secretion:** Vesicles carrying most secretory proteins leave the TGN, in a continuous process fuse with the nearby plasma membrane, and release the contents of the vesicle to

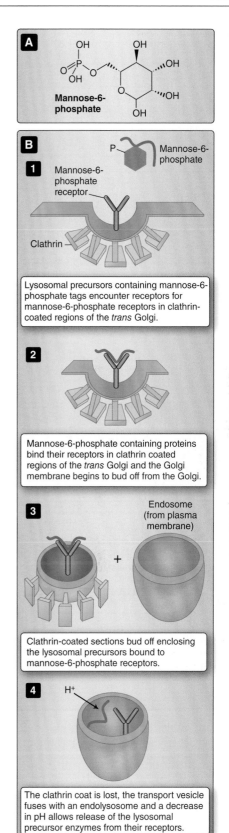

A Mannose-6-phosphate

B

1 Mannose-6-phosphate receptor / Mannose-6-phosphate / Clathrin

Lysosomal precursors containing mannose-6-phosphate tags encounter receptors for mannose-6-phosphate receptors in clathrin-coated regions of the *trans* Golgi.

2 Mannose-6-phosphate containing proteins bind their receptors in clathrin coated regions of the *trans* Golgi and the Golgi membrane begins to bud off from the Golgi.

3 Endosome (from plasma membrane)

Clathrin-coated sections bud off enclosing the lysosomal precursors bound to mannose-6-phosphate receptors.

4 H⁺

The clathrin coat is lost, the transport vesicle fuses with an endolysosome and a decrease in pH allows release of the lysosomal precursor enzymes from their receptors.

Figure 11.8
Movement of new proteins to lysosomes.

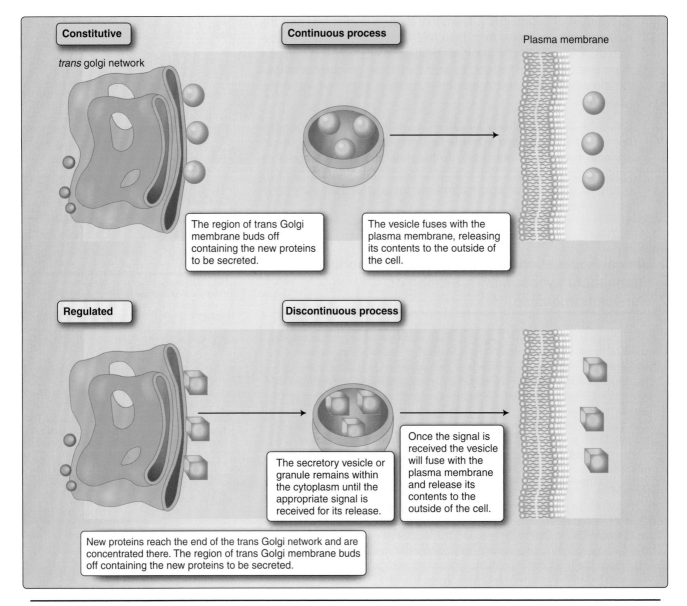

Figure 11.9
Constitutive and regulated secretions.

the outside of the cell (Figure 11.9). This process is known as constitutive secretion and it operates for proteins released regularly from the cell that produced them. Extracellular matrix proteins, including collagen, elastin, and fibronectin, are examples of proteins that are secreted in a constitutive manner from cells within connective tissues (see also Chapter 2).

b. **Regulated secretion:** Other proteins are released from cells only at certain times, in a discontinuous process known as regulated secretion or **exocytosis**. Proteins released in this manner generally play important regulatory roles. They are concentrated within the TGN prior to their release within a transport

A. General mechanism of lysosomal protein degradation

Lysosomes are membrane-enclosed organelles that contain digestive enzymes including those that function as lipases, nucleases, and proteases. The interior of the lysosome is more acidic than the cytosol (pH 4.8 vs. 7.2). This compartmentalization or segregation of lysosomal enzymes is important to prevent uncontrolled degradation of functional cellular contents by the actions of these potent digestive enzymes. One pathway for the uptake of molecules targeted for degradation involves **autophagy**, a process by which vesicles are formed (autophagosomes) that engulf small amounts of cytoplasm or specific organelles using portions of the endoplasmic reticulum (Figure 12.2). Fusion of the vesicles with lysosomes results in the release of lysosomal hydrolytic enzymes allowing for degradation of the macromolecules. While different autophagic pathways exist, causing both selective and nonselective protein degradation, they also share a number of common steps making these pathways both specific and flexible.

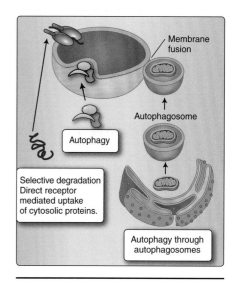

Figure 12.2
General scheme of lysosomal protein degradation.

Clinical Application 12.1: Autophagy and Neurodegenerative Diseases

In chronic degenerative diseases such as Huntington, Alzheimer, and Parkinson diseases, there is an abnormal accumulation of defective proteins within nervous tissue. Because it was demonstrated that autophagosomes accumulate in brains of patients with these diseases, it had been thought that autophagy contributed to the pathogenesis of these disorders. However, recent more compelling evidence suggests that autophagy may actually work to protect against diverse neurodegenerative diseases. And the accumulation of autophagosomes is now thought to primarily represent the activation of autophagy as a beneficial physiological response in these pathological conditions.

Selective lysosomal degradation

Under certain conditions, lysosomes selectively degrade cytosolic proteins. One example of selective degradation occurs during starvation, when proteins containing the amino acid sequence Lys-Phe-Glu-Arg-Gln are targeted to the lysosomes. This process also requires unfolding of the polypeptide chains of the protein (mediated by cytosolic chaperones) and a lysosomal membrane receptor to transport proteins across the lysosomal membrane. The target proteins usually have long half-lives and tend to be dispensable proteins that under conditions of stress and starvation are sacrificed to release amino acids and produce energy to support basic metabolic reactions (see Figure 12.2).

B. Proteosomal degradation

The ATP-dependent proteosomal pathway involves the protein **ubiquitin.** This is a highly conserved protein containing 76 amino acids that, as its name suggests, is ubiquitous in the eukaryotic kingdom. Proteins destined for destruction in the proteosomal pathway are tagged by

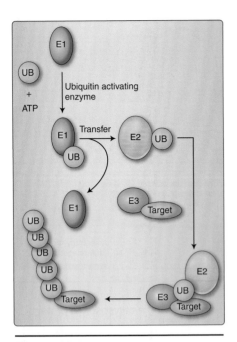

Figure 12.3
Steps in protein ubiquitination.

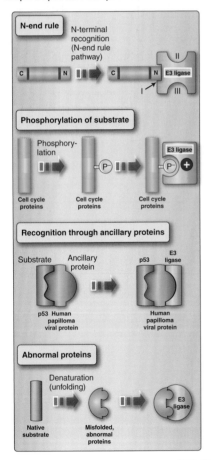

Figure 12.4
Modes of recognition of substrates for degradation.

covalent attachment of ubiquitin and then subsequently degraded in a proteolytic complex called the **proteosome** (see Figure 12.1).

1. **Ubiquitination of proteins:** Ubiquitination is a highly regulated process that is critically important for protein degradation. Ubiquitin is covalently linked to proteins via an ATP-dependent pathway involving three separate enzymes E1, E2, and E3 (Figure 12.3). These reactions result in linking the carboxy-terminal glycine residue in ubiquitin to a lysyl residue in the protein to be degraded. The process occurs through a sequence of three steps requiring the ubiquitin-activating enzyme, E1, to initially bind ubiquitin followed by the transfer to the ubiquitin-conjugating enzyme, E2. The ubiquitin ligase, E3, then promotes transfer of ubiquitin from E2 to the lysyl residue of the protein recognized by the E3 as being slated for degradation. Generally, proteins become polyubiquitinated, with several ubiquitin molecules added, forming a chain of ubiquitins linking a lysyl residue in one ubiquitin molecule to the carboxy-terminus of the adjacent ubiquitin. A minimum of four ubiquitin molecules on the target protein appears to be critical for efficient degradation of that protein.

2. **Modes of recognition of substrates for degradation:** The half-lives of proteins correlate with their amino-terminal residue. In general, proteins with an N-terminal Met, Ser, Ala, Thr, Val, or Gly have half-lives greater than 20 hours and proteins with N-terminal Phe, Leu, Asp, Lys, or Arg have half-lives of 3 minutes or less. Proteins that are rich in Pro (P), Glu (E), Ser (S), and Thr (T) ("PEST" proteins) are more rapidly degraded than other proteins. Other mechanisms for recognition include recognition of phosphorylated substrates, recognition of ancillary proteins bound to the substrate, and recognition of mutated abnormal proteins. Different classes of enzymes (E3 ligases—see below) are involved in each pathway for degrading specific substrates as shown in Figure 12.4.

3. **Proteosome:** The proteosome is a large, 26S complex of proteins, made up of approximately 60 protein subunits, and structurally resembles a large cylinder that is covered on both ends (Figure 12.5). It contains a central 20S core and a 19S regulatory particle at either end. The central 20S core particle is barrel shaped and formed of four rings. The outer rings are composed of seven alpha subunits and the inner ring is formed from seven beta subunits. Some of the beta subunits have protease activity.

The 19S regulatory particle is important for several activities, including recognition and binding of polyubiquitinated proteins, removal of ubiquitin, unfolding the protein substrate, and translocation into the central core. These diverse functions are facilitated by the complex composition of the 19S particles, made up of several ATPases and other enzymes. The unfolded proteins are then hydrolyzed within the central core into smaller peptides. The smaller peptides emerge from the opposite end of the 20S particle and are further degraded by cytosolic peptidases (Figure 12.5).

Clinical Application 12.2: Cancer Causing HPVs Target Host Cellular Proteins for Degradation

It is widely accepted that certain human papillomaviruses (HPVs), including HPV types 16 and 18, play an etiologic role in cervical carcinogenesis. The major oncoproteins of these HPVs are encoded by the E6 and E7 genes, which are the only viral genes that are generally retained and expressed in HPV-positive cancer cells. P53 tumor suppressor protein (see Chapters 21 and 22) is a target of these high-risk HPV strains. However, unlike most human cancers where p53 is inactivated by missense mutations, the mechanism of inactivation in cervical cancers is unique. P53 is targeted by the HPV E6 oncoprotein, which binds it and utilizes the cell's ubiquitin-protein ligase E6-AP to target p53 for degradation (see Figure 12.3). In normal cells, p53 is generally targeted for degradation by a different ubiquitin ligase called Mdm2 (an E3 ligase) rather than E6-AP. While some classes of E3 proteins function as adaptors that bring E2 enzymes into complex with their substrates (the RING class of proteins to which Mdm2 belongs), E6-AP belongs to a class of ubiquitin-protein ligases called HECT E3 proteins, which directly transfer ubiquitin to their substrates.

Clinical Application 12.3: Proteosome Inhibitors as Anticancer Agents

Bortezomib is the first therapeutic proteosome inhibitor to be tested in humans. It is approved for the treatment of multiple myeloma (cancer of the plasma cells) and mantle cell lymphoma (a rare cancer of the lymphocytes). The drug is a peptide and binds in the catalytic site of the 26S proteosome and inhibits the degradation of proteins. Bortezomib is thought to inhibit the degradation of proapoptotic factors, thereby increasing the death of the cancer cells through apoptosis. Other mechanisms may exist for the effectiveness of the drug.

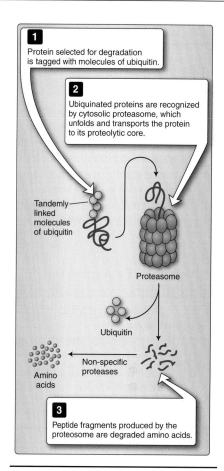

1 Protein selected for degradation is tagged with molecules of ubiquitin.

2 Ubiquinated proteins are recognized by cytosolic proteasome, which unfolds and transports the protein to its proteolytic core.

Tandemly linked molecules of ubiquitin

Proteasome

Ubiquitin

Non-specific proteases

Amino acids

3 Peptide fragments produced by the proteosome are degraded amino acids.

Figure 12.5
Degradation of proteins in proteosomes.

Chapter Summary

- All proteins are degraded eventually using the cell's proteolytic system.

- The two systems for protein degradation are the lysosomal protein degradation pathway and the ATP-dependent proteosomal degradation pathway.

- Proteins with shorter half-lives are degraded through the proteosomal pathway, while proteins with longer half-lives utilize the lysosomal pathway.

- Autophagy through the lysosomal pathway is important for generation of energy and amino acids during cellular stress.

- Proteins destined for degradation are covalently attached to a chain of ubiquitin residues.

- Proteosomal degradation is initiated by a multistep process requiring enzymes that add ubiquitin to the protein targeted for degradation.

- Proteosomes are large protein complexes that carry out the actual breakdown of proteins to smaller peptides while regenerating ubiquitin.

- The half-life of the protein is determined both by the amino-terminus residue and by the amino acid composition of the protein.

Study Questions

Choose the ONE best answer.

12.1 Autophagy refers to

    A. Removal and subsequent breakdown of membrane vesicles within the cells.

    B. Breakdown of cytoplasmic proteins in the lysosomal compartment.

    C. A process that generates energy and amino acids when a cell is under stress.

    D. The formation of an autophagosome followed by digestion by lysosomal hydrolases.

    E. All of the above.

> Correct answer = E. Autophagy is a process by which organelles or cytoplasmic proteins are degraded in the lysosomal compartment. This process usually proceeds through the formation of an autophagosome, which then fuses with the lysosomal membrane causing the release and degradation of the contents. Autophagy is also activated when the cell is under stress and requires raw materials such as amino acids and energy.

12.2 A protein with a short half-life

    A. Usually has an N-terminus serine amino acid.

    B. Is preferentially degraded by the lysosomal pathway.

    C. Has a C-terminal phenylalanine amino acid.

    D. Is tagged with ubiquitin before degradation.

    E. Is degraded into constituent amino acids in autophagosomes.

> Correct answer = D. Proteins with short half-lives are usually degraded through the proteosome pathway after tagging them with ubiquitin. Proteins with N-terminus serine have a long half-life and are preferentially degraded by the lysosomes. The C-terminal amino acid residue does not affect the half-life of a protein. Autophagosomes are intermediates in the lysosomal protein degradation pathway.

12.3 A proteosome is a(n)

    A. Proteolytic complex that degrades all cellular proteins.

    B. Enzyme complex required for the addition of ubiquitin to proteins destined for degradation.

    C. Proteolytic complex consisting of ATPases and other enzymes to degrade proteins.

    D. Complex composed of a central core and one regulatory particle.

    E. Complex of proteins that is present in the lysosomes of cells.

> Correct answer = C. Proteosomes are barrel-like structures that are made of several protein subunits with an ability to degrade intracellular ubiquitinated proteins. This process requires ATP. It selectively degrades intracellular proteins with short half-lives, and these proteins have to be ubiquitinated. Proteosomes remove ubiquitin from the target proteins and are composed of a central core and two regulatory regions. Proteosomes are present in the cytosol of cells and are distinct from the lysosomal pathway of protein degradation.

12.4 A nonfunctional intracellular protein that is rich in PEST amino acid residues and has an N-terminal Phe residue will most likely be degraded via

    A. Autophagy.

    B. Peroxisomal digestion.

    C. Acid hydrolase action.

    D. Phagocytosis.

    E. An ATP-dependent pathway.

> Correct answer = E. An ATP-dependent process, using proteasomes, is used to degrade proteins rich in PEST residues. Additionally, those proteins with N-terminal Phe residues have short half-lives and are degraded by proteasomes. Lysosomes use acid hydrolases to digest macromolecules in a process involving autophagy. Lysosomes generally degrade proteins with long half-lives, and not PEST-containing proteins with short half-lives. Peroxisomes do not degrade large macromolecules like proteins. Instead, they detoxify hydrogen peroxide and break down fatty acids and purines (see Chapter 5, pp. 52, 53). Phagocytosis is a process by which vesicles are internalized within cells. If the contents of the vesicle are to be degraded, lysosomal digestion is used.

12.5 A 6-month-old, previously healthy male child begins to lose motor skills and is diagnosed with Tay-Sachs disease. Over the next year, he experiences more and more accumulation of gangliosides in his brain, which contributes to worsening of his signs and symptoms. A defect in which of the following caused this child's condition?

- A. An acid hydrolase
- B. Peroxisomal digestion
- C. Ubiquitination
- D. An ATP-dependent pathway
- E. PEST-containing proteins

Correct answer = A. Infantile Tay-Sachs disease results from a mutation in the *HEXA* gene, which encodes beta hexosaminidase A, an acid hydrolase that normally functions within lysosomes. Tay-Sachs is a lysosomal storage disease (see also p. 52). Gangliosides are glycosphingolipids, not proteins, and are therefore not degraded via the ATP-dependent, ubiquitin tagging system of proteasomes that do act to degrade PEST-containing and many other proteins. Peroxisomes act to detoxify hydrogen peroxide and do break down fatty acids and purines, but not the gangliosides that accumulate in individuals with Tay-Sachs disease.

*There is a point where in the mystery of existence
contradictions meet; where movement is not all movement
and stillness is not all stillness; where the idea and the form,
the within and the without, are united; where infinite becomes
finite, yet not*

—Rabindranath Tagore (Indian poet, 1861–1941)

Oftentimes in living cells, the concentration of a critical molecule is greater outside than within the confines of the plasma membrane. Ions and nutrients are often able to cross the barrier, nourishing and providing the cell with essential constituents. Such molecules generally move along or down the concentration gradient of the molecule in a passive manner. In the first chapter of this unit, we will consider basic concepts of transport, mainly focusing on passive transport processes. Additionally, we will consider osmosis, or movement of water. Since cells exist within an aqueous environment, water, the fluid of life, continuously flows into and out of cells via protein channels in the membrane. While water transport is constant and involves large volumes over time, there is no net movement of water when isotonic conditions exist. The vast movement appears to be negligible when it is not! At times, a cell requires a crucial molecule that is already present in a higher concentration within than outside its boundaries. Contrary to what may be viewed as a prudent use of energy currency, a cell may use an active transport process, powered by ATP to pump in the molecule, which is then sometimes used to generate more energy for the cell. Active transport is the topic of the second chapter in this unit. For our third chapter, we discuss in detail the transport of glucose into cells as cells have an almost insatiable need for this carbohydrate. Disruptions of glucose transport are seen in diabetes mellitus. The final chapter in this unit concerns transport of drugs. These synthetic molecules can exploit the transport mechanisms that have evolved to meet normal cellular needs. With them, we have treatments for many illnesses and perhaps even the potential for extending the finite existence of our cells and our species.

13 Basic Concepts of Transport

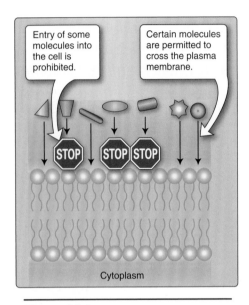

Figure 13.1
Selective permeability of the plasma membrane.

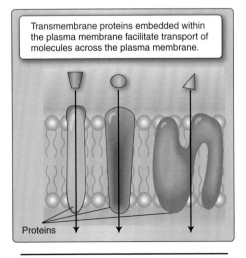

Figure 13.2
Membrane proteins facilitate transport.

I. OVERVIEW

Certain ions and molecules must enter and exit cells. Cells must maintain their volume and a balance of ions in order for normal processes of cellular life to occur. Molecules that serve as cellular food must also enter cells so that they can be broken down for energy. The plasma membrane protects and isolates the cytoplasm from the environment and, in so doing, presents a barrier to the entry of molecules into the cell. This barrier is **selectively permeable**, allowing physiologically important molecules to enter and exit cells while excluding other molecules (Figure 13.1). Some drugs can also often gain access to the interior of the cell.

Proteins embedded in the plasma membrane are important in facilitating transport of ions and nutrients into cells (Figure 13.2). These proteins are **ion channels, transporters**, and pumps. Individual membrane proteins specifically bind to certain ligands (e.g., to chloride, to glucose, or to sodium and potassium) and facilitate their movement across the plasma membrane.

In most membrane transport, **passive transport** is used where molecules are moved across the plasma membrane in the direction of their concentration gradients (Figure 13.3). The molecule flows from where it is at higher concentration to where it is at a lower concentration. In contrast, in **active transport**, molecules are moved against their concentration gradient in energy-requiring processes (see also Chapter 14).

II. DIFFUSION

Knowledge of diffusion is useful in describing the movement of molecules from higher concentration to lower concentration. Diffusion is powered by the random movement of molecules in a solution, where molecules spread out until they are equally distributed within the space they occupy (Figure 13.4). The process can continue at the same rate as long as the concentration gradient exists. Diffusion of one substance does not interfere with the diffusion of another substance within the same solution. The net movement or flux of a substance diffusing across a barrier depends on several criteria. First is the concentration gradient, next is size of the molecule, and then the permeability of the substance in the barrier across which it will diffuse.

A. Considerations

Permeability in a membrane is an important consideration if a molecule is to cross it by diffusion. Phospholipids that constitute the plasma membrane are **amphipathic** in nature, containing both hydrophilic (water-loving) and hydrophobic (water-fearing) components (see also Chapter 3). Hydrophobic molecules such as steroid hormones may be able to dissolve in the interior hydrophobic core of a plasma membrane. However, the hydrophilic head groups of the phospholipids present an obstacle at the interface with the environment and with the cytosol (Figure 13.5).

B. Diffusion and plasma membranes

In sheets of cells in tissues, some molecules may diffuse through tight junctions between neighboring cells (see also Chapter 2). In this way, some molecules diffuse through the layer of cells, but not into the cytoplasm of cells. *The plasma membrane is always a barrier to diffusion* and impedes the flow of materials through it. Consequently, ordinary, unaided diffusion of most molecules across plasma membranes does not normally occur. Movement of molecules along their concentration gradient into the cytoplasm of cells occurs through membrane proteins that form channels within the plasma membrane. Some ions and small molecules (generally with a molecular weight of <80 daltons), including some gases, may gain access to the cytoplasm of cells without a specific transport protein of their own, by using protein channels that exist in plasma membranes for the transport of water (aquaporins). Larger molecules require specific membrane transport proteins to cross the boundary of the plasma membrane. Water entry into cells by osmosis has been described as diffusion, but it too occurs via membrane channels.

III. OSMOSIS

Osmosis is the transfer of a liquid solvent through a semipermeable membrane that does not allow the passage of certain dissolved solutes. Water is the most important physiological solvent. Plasma membranes are permeable to water but not to some solutes in solution in the water. Protein channels known as **aquaporins** enable water to pass through the hydrophobic core of the phospholipids of the membrane. By osmosis, water crosses a plasma membrane from an area of high concentration of water (lower concentration of solute) to an area of lower concentration of water (higher concentration of solute) (Figure 13.6).

A. Net movement of water

Water constantly enters and exits our cells by osmosis. However, net movement of water is usually negligible. Rates of water entry and exit are usually equal. In erythrocytes, water equal to approximately 250 times the volume of the cell enters and exits the cell each second, but with no net movement of water! In the small intestine though, there is a net movement of water across sheets of cells as water is absorbed and secreted. Also, net movement of water via osmosis occurs to concentrate the urine. In this instance, water moves from the filtrate that will form urine across a layer of epithelial cells lining the kidney tubules, and into the blood.

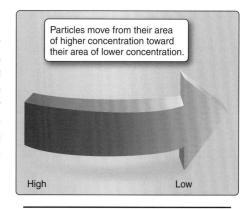

Figure 13.3
Concentration gradients of solutes to be transported.

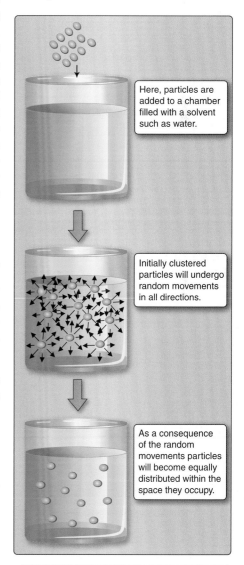

Here, particles are added to a chamber filled with a solvent such as water.

Initially clustered particles will undergo random movements in all directions.

As a consequence of the random movements particles will become equally distributed within the space they occupy.

Figure 13.4
Distribution of particles in solution via diffusion.

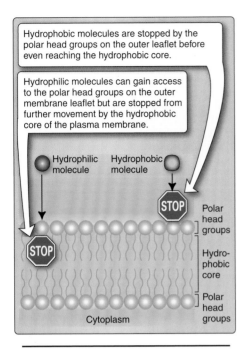

Figure 13.5
Amphipathic phospholipids as barriers to membrane diffusion.

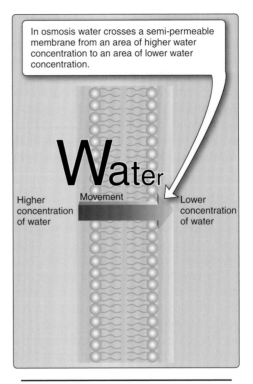

Figure 13.6
Osmosis.

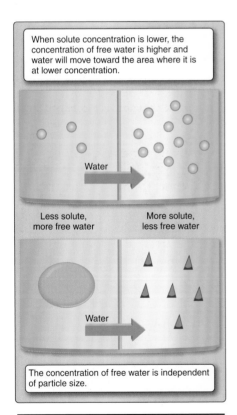

Figure 13.7
Concentrations of free water determine the direction of water movement in osmosis.

Solute concentrations will determine the concentration of free water. In a solution with a high concentration of particles or solute, there is less free water than in a solution with low solute concentration (Figure 13.7). Where there is more free water, the water molecules will strike the aquaporins in the membrane with greater frequency and more water will leave the area of high free water concentration. The result is net movement of water. The size or molecular weight of the solutes in water does not influence the net movement of water. When sufficient water has crossed the membrane to equalize solute concentrations on both sides, then movement of water will stop.

B. Osmotic pressure

Differences in solute concentration on opposing sides of a barrier such as a plasma membrane generate an osmotic pressure. If the pressure is raised on the side of the barrier into which water is flowing, movement of water will stop (Figure 13.8). This is referred to as the **hydrostatic** or water-stopping pressure.

C. Cell volume

When the osmotic pressure is the same both inside and outside the cell, the external solution surrounding the cell is considered to be **isotonic** (same). Cells maintain their volume in isotonic solutions. While osmosis occurs both into and out of the cell, there is no net movement of water when osmotic pressure is equal on both sides of the membrane (Figure 13.9). A solution with less osmotic pressure than the cytosol is **hypotonic**. Cell volume increases in hypotonic solutions.

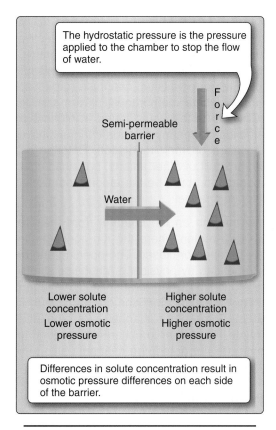

Figure 13.8
Osmotic pressure generated by differences in solute concentration on different sides of a barrier.

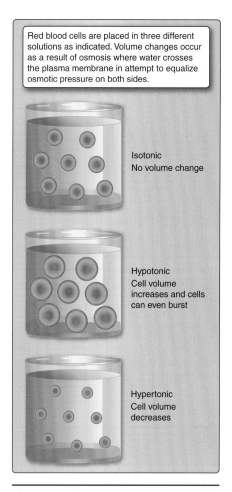

Figure 13.9
Cell volume changes as a consequence of osmosis.

This occurs because free water is at higher concentration outside the cell. Water moves with or along its concentration gradient and there is a net movement of water into cells. If cells are instead placed in a solution with a higher osmotic pressure than their cytosol, the solution is **hypertonic.** Water will exit the cells in this attempt to equalize osmotic pressure, and cell volume will decrease.

IV. PASSIVE TRANSPORT

Based on their size, charge, and low solubility in phospholipids, many biologically important molecules such as ions, sugars, and amino acids would be predicted to enter into cells very slowly. However, their uptake into cells can occur at fairly rapid rates (Figure 13.10) because **ion channels** or **membrane transport proteins** facilitate the movement of specific molecules into and out of cells. This process is also called **catalyzed transport** and sometimes referred to as facilitated diffusion. No direct source of energy is required to power the process.

A. Transport through ion channels

Ions enter cells through ion channels, complexes of transmembrane proteins within the plasma membrane that provide a hydrophilic

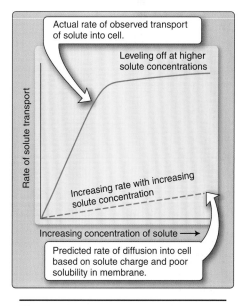

Figure 13.10
Rates of transport into cells.

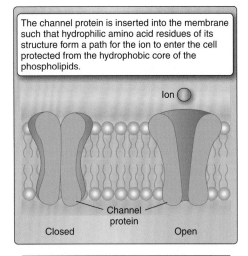

The channel protein is inserted into the membrane such that hydrophilic amino acid residues of its structure form a path for the ion to enter the cell protected from the hydrophobic core of the phospholipids.

Ion

Channel protein

Closed   Open

Figure 13.11
Ion channels.

route for an ion to pass through the hydrophobic core of the plasma membrane (Figure 13.11). Ion channels are selective, allowing ions of only certain size and charge to enter. For example, calcium channels are specific for calcium and sodium channels for sodium. While many distinct types of potassium channels (grouped into four major classes) exist in plasma membranes, all are specific for uptake of potassium. Chloride is the anion found in highest concentration in physiological fluids. The many chloride channels have been grouped into families, based on their regulation. Chloride channels may at times permit transport of other anions, but since chloride is found in highest concentration, it will be the anion most likely transported by any given chloride channel.

All ions move through an ion channel from their region of higher concentration to a region where the ion concentration is lower. Ion channels may be open or closed. The opening and closing of **gated channels** is regulated by physical or chemical stimuli. Ligand-gated channels are regulated by neurotransmitters. Voltage-gated channels are regulated by an electric field and are particularly important in the nervous system. Some other channels are regulated by G proteins (see also Chapter 17). Other ion channels respond to osmotic pressure and some are not gated. Ion channels also have properties in common with transporters, described below.

B. Transport through transporters

Transporters are transmembrane proteins that catalyze migration of molecules, referred to as ligands, across plasma membranes. **Uniport** is the facilitated transport of one molecule by a transporter known as a uniporter. Transporters or uniporters are also sometimes called **permeases** because of their enzyme-like function in catalyzing movement of their ligand across the plasma membrane. Also, because of the similarities between membrane transport and enzyme-catalyzed reactions, the ligand being transported is often called the substrate of the transporter. Just as enzymes exhibit specificity toward certain substrates, individual transporters are able to bind to and interact with only certain ligands (see also *LIR Biochemistry*, Chapter 5, for a discussion of enzymes). The major physiological form of glucose, D-glucose, has a much higher affinity for its transport protein than does L-glucose, a stereoisomer of glucose (see also Chapter 15).

Clinical Application 13.1: Cystic Fibrosis and Defective Chloride Ion Transport

Cystic fibrosis (CF) is the most common lethal genetic disease in Caucasians, with a prevalence of about 1:2,500 births. CF is also common in the Ashkenazi Jewish population but is rare in African and Asian populations. Certain mutations of the cystic fibrosis transmembrane conductance regulator (CFTR) cause the disease. CF is inherited as an autosomal recessive trait, with approximately 1 in 25 individuals of Caucasian descent being a carrier, with one copy of a mutant *CFTR* gene. Inheritance of two mutant *CFTR* genes is necessary for development of CF. While over 2,000 mutations are known in *CFTR*, most do not cause disease. Some disease-causing mutations result in more severe forms of disease than do others.

Clinical Application 13.1: Cystic Fibrosis and Defective Chloride Ion Transport (Continued)

Normal CFTR functions as a chloride channel in epithelial cells. Defective chloride ion transport occurs when two copies of disease-causing mutations in *CFTR* are present. A common, severe mutation in *CFTR*, ΔF508 (deletion of 3 base pairs and loss of a phenylalanine), impacts folding of the CFTR protein. In persons with two copies of this mutation (homozygous), the defective CFTR protein never inserts into the plasma membrane. Having a "salty brow" was an early diagnostic test for CF. The sweat of affected individuals contains more salt than normal, as a result of inappropriate chloride ion transport. Chloride sweat tests have been used in diagnosis of CF.

Defective CFTR has much more serious consequences than causing salty sweat. Inability to transport chloride across epithelial cells leads to decreased secretion of chloride and increased reabsorption of sodium and water, resulting in the production of thick sticky secretions in the lungs and increased susceptibility to infections. The leading cause of death of persons with CF is respiratory failure often during their 20s or 30s. In the pancreas, thick mucus often prevents pancreatic enzymes from reaching the intestine to aid in digestion of dietary lipids. Most males with CF experience infertility because mutation of *CFTR* also usually results in the absence of the vas deferens, which is needed for release of the sperm.

Treatments for CF include percussion therapy to loosen respiratory mucus, antibiotics for infections, and pancreatic enzyme replacement therapy. With improved treatments, affected individuals may survive into their 40s or 50s, but no cure is available at present. Gene therapy continues to be a future possibility.

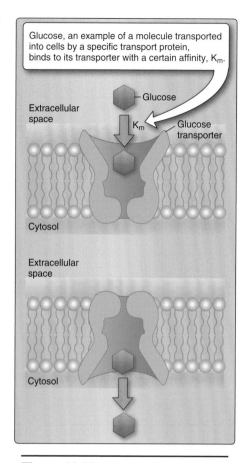

Figure 13.12
Transport proteins.

Catalyzed transport via transporters (and ion channels) requires a concentration gradient of the solute or the substrate of the transporter. Molecules bind to their specific transporter with a certain **affinity**, represented as the K_m (Figure 13.12) (see also *LIR Biochemistry*, Chapter 5, for discussion of affinity of enzyme for substrate, K_m, which is analogous to affinity of transporter and solute). Numerically, K_m is equal to the concentration of solute that yields half the maximal velocity of transport. This **maximal velocity** (V_{max}) of transport will be achieved when all available transporter proteins are bound by specific solute or substrate. After that point, the addition of more solute/substrate will not result in an increased rate of uptake into the cells. Therefore, membrane transport via transporters (and via ion channels) is a **saturable** process (Figure 13.13).

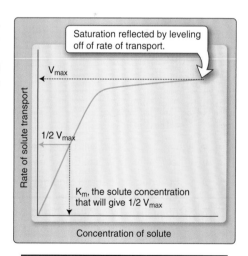

Figure 13.13
Characteristics of transport catalyzed by transporters.

Chapter Summary

- Selective permeability of the plasma membrane allows certain materials only to enter and exit cells.
- Diffusion of particles is powered by movement of the particles within a solution and results in a distribution of particles that move from where they are at higher concentration to an area where they were at lower concentration.
- Many molecules enter (or exit) cells by moving from an area of high concentration to low concentration.
- The plasma membrane is always a barrier for a substance to access the cytoplasm; entry into cells requires membrane proteins.
- Water enters and exits cells through osmosis and requires water channel proteins called aquaporins.
- Ion channels facilitate movement specific ions into or out of cells, along the concentration gradient of the ions.
- Transport proteins catalyze the movement of specific solutes or substrates across plasma membranes by passive transport, also called catalyzed transport and facilitated diffusion.
- Ion channels and transport proteins bind their solute/substrate with a certain affinity and catalyze the movement of that solute across the membrane in a saturable manner.

Study Questions

Choose the ONE best answer.

13.1 A small volume of water containing a high concentration of particle C is added to a container of water that already contains particles A and B, as shown. Particles A, B, and C are equally soluble in water.

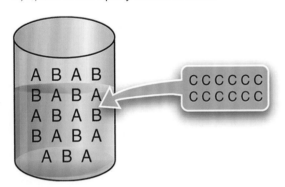

Which of the following will occur after the addition of the water containing particle C?

A. Clustering of A and B together to occupy less space within the container
B. Competition by particles A and B with particle C for space in the container
C. Equal distribution of C throughout the container without regard to A and B
D. Movement of all C particles to the water beneath where A and B are found
E. Segregation of particle C to the topmost portion of the container only

Correct answer = C. Particle C will move by diffusion to be equally distributed throughout the solution without regard to particles A or B. No competition or clustering will occur. Since all three types of particles are equally soluble in water, random movement of all three will occur, resulting in their equal distribution within the container.

13.2 Two different aqueous solutions are placed in equal-sized chambers on either side of a semipermeable membrane as shown. The membrane is impermeable to the particles but permeable to water. Which of the following will occur?

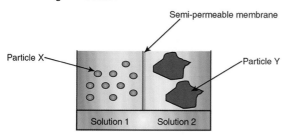

A. No net movement of water or of solutes will occur.
B. One molecule of particle Y will move to solution 1.
C. Osmosis will cause water to exit solution 1.
D. Particle X will move from solution 1 to solution 2.
E. Water will leave solution 2 and enter solution 1.

Correct answer = E. Water will leave solution 2 in an attempt to equalize the osmotic pressure on both sides of the barrier that is permeable to water. Particle size has no bearing on the process. More free water exits in solution 2 than in solution 1. Neither particle X nor particle Y can cross the membrane since it is impermeable to both solutes. Since water moves by osmosis from its area of higher to lower concentration, it cannot move from solution 1 to solution 2 because solution 1 has less free water than does solution 2. There is net movement of water in this process.

13.3 Erythrocytes are placed in a hypertonic solution of sodium chloride. Which of the following will be an effect of the osmosis that will occur?

A. Cells will burst.
B. Cells will shrink in volume.
C. Net movement of sodium chloride will occur into the cells.
D. Osmotic pressure will decrease inside the cells.
E. Water will enter into the cells.

Correct answer = B. Cells reduce their volume when placed in hypertonic solutions. In such an instance, there is more free water inside cells than outside cells. Water moves via osmosis out of the cells, reducing the cell volume. Cells do not burst, which could occur in hypotonic solutions where there is net movement of water into cells by osmosis. Sodium chloride does not move by osmosis, which is the movement of water through plasma membranes. The osmotic pressure within the erythrocytes will increase and not decrease as a result of loss of water, which will effectively increase the internal concentration of sodium chloride.

13.4 Cells are placed in an environment where the concentration of extracellular sodium is greater than the intracellular concentration of sodium but the extracellular concentration of calcium is less than the intracellular concentration of calcium. Which of the following will occur?

A. Calcium movement outside the cell to balance the sodium concentration
B. Calcium competing with sodium binding to sodium ion channels outside the cell
C. Diffusion of sodium directly through the hydrophobic core of the plasma membrane
D. Sodium binding to a sodium ion channel and being transported along its gradient
E. Sodium preventing calcium from binding to calcium channels outside the cell

Correct answer = D. Sodium is at higher concentration outside the cell than inside the cell. Sodium ion channels will facilitate its movement along its concentration gradient to the inside of the cell. Calcium is at higher concentration inside the cell than outside, so it will not move against its concentration gradient. Sodium will have a much higher affinity for a sodium ion channel than calcium, and calcium will not compete with sodium, particularly when calcium does not have a strong concentration gradient. Since sodium is a charged particle, it cannot simply diffuse unaided through the hydrophobic core of the plasma membrane. No concentration gradient exists for calcium from outside to inside the cell; therefore, the presence of sodium will not interfere with the working of calcium channels designed to transport calcium into cells.

13.5 Transport of glutamine through a glutamine transporter into certain cells is observed to be 0.5 picomole (10^{-12} mole) per million cells per second when the extracellular concentration of glutamine is 150 micromolar (10^{-6} moles/L) and 1 picomole per million cells per second when the concentration of glutamine is 3,000 micromolar. However, the rate of transport is also 1 picomole per million cells per second when the concentration of glutamine is 3,500 micromolar, 4,000 micromolar, and 6,000 micromolar. These data demonstrate

A. Lack of a concentration gradient for glutamine.
B. Low affinity of glutamine for its transporter.
C. Nonspecificity of the glutamine transporter.
D. Saturation of the glutamine transporter with glutamine.
E. That V_{max} cannot be achieved under these conditions.

Correct answer = D. The glutamine receptor is saturated with glutamine and cannot transport glutamine any faster than the V_{max} of 1 picomole per million cells per second that has already been achieved. A concentration gradient must exist in order for glutamine to be taken up by its transporter. Since saturable transport is occurring, the transporter must have specificity for glutamine, with sufficiently high affinity that transport is occurring. V_{max} has been achieved. It is the maximal velocity of transport, 1 picomole per million cells per second.

13.6 In passive transport via uniport, substrates are transported

A. Along, with, or down their concentration gradients.
B. Independent of transmembrane proteins.
C. Less rapidly than predicted by partition coefficient and size.
D. Two at a time: one into the cell and one out of the cell.
E. Using ATP hydrolysis to power the movement.

Correct answer = A. Passive transport via uniport moves substrates along, with, or down their concentration gradients. Transmembrane proteins are required for this form of transport, which occurs more rapidly than predicted by partition coefficient and size. One substrate is transported at a time via uniport. Energy from ATP hydrolysis is not needed.

13.7 Results of neonatal screening indicate a diagnosis of cystic fibrosis (CF) for a female child. The signs and symptoms of this disease include frequent respiratory infections and reduced ability to digest fats and result from defective

A. Elastin synthesis in lungs.
B. Immune cell surveillance.
C. Production of pancreatic enzymes.
D. Release of sweat.
E. Transport of chloride ions.

Correct answer = E. The signs and symptoms of CF develop in response to impaired transport of chloride ions, owing to inheritance of two mutations in the *CFTR* gene. While lung and pancreatic involvement are common in most patients with CF, lung elastin is normal and pancreatic enzymes are produced. However, owing to inappropriate chloride ion transport, mucus becomes thick and is trapped in lungs, facilitating bacterial infections. Thick mucus prevents secretion of pancreatic enzymes that are needed to digest lipids. The immune surveillance in patients with CF is not altered. But infections result from the thick mucus. Sweat in individuals with CF has a higher salt content, owing to defective chloride ion transport, but sweat is released from the glands. The salt that reaches the surface of the skin in the sweat is not reabsorbed when both copies of *CFTR* are defective.

Active Transport 14

I. OVERVIEW

Active transport occurs when molecules or ions are moved across cell membranes *against* their concentration gradients (Figure 14.1). Energy is required to move these solutes in the opposite direction of their concentration gradient. The energy is derived from the hydrolysis of adenosine triphosphate (**ATP**). The membrane proteins that bind to and transport the substrates against their gradients possess the ATPase enzymatic activity to directly hydrolyze ATP to yield adenosine diphosphate (ADP) and inorganic phosphate (P_i) to harness its energy. These ATP-powered pumps function in **primary active transport**.

A consequence of primary active transport is the establishment of **ion gradients**. Certain ions such as sodium are pumped out of cells while others, such as potassium, are pumped into cells by primary active transport. Sodium is therefore found at much higher concentration outside of cells than inside of cells and potassium is at higher concentration inside cells than outside as a consequence of primary active transport. The tendency of these ions will be to move down their concentration gradients. Such strong concentration gradients, like those for sodium, can be used to power the transport of other solutes. By binding to specific membrane transport proteins that also bind to the ion (cotransport), these other molecules can travel along the concentration gradient of the ion, along with the ion, even against the direction of their own concentration gradient. **Secondary active transport** is the process by which ion gradients generated by ATP-powered pumps are used to provide the energy for transport of other molecules and ions against their own concentration gradients.

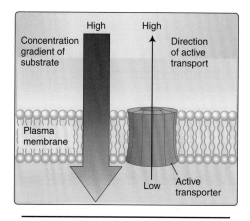

Figure 14.1
Active transport moves substrates against their concentration gradients.

II. PRIMARY ACTIVE TRANSPORT

Four classes of transport proteins function as ATP-powered pumps to transport ions and molecules against their concentration gradients (Figure 14.2). All have ATP-binding sites on the cytosolic side of the membrane. ATP hydrolysis is coupled to the transport of the substrates of the ATP-powered pump. ATP is only hydrolyzed to ADP + P_i when particular ions or molecules are transported. The classes differ in the types of ions/molecules transported and in the mechanisms used to catalyze ATP-powered transport.

Class	Substrate(s) transported
P	Ions (H^+, Na^+, K^+, Ca^{2+})
F	H^+ only
V	H^+ only
ABC	Ions, drugs, xenobiotics

Figure 14.2
Four classes of primary active transporters.

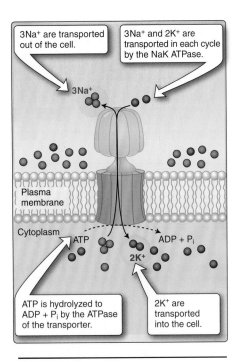

Figure 14.3
The sodium-potassium ATPase pump.

A. P-class pumps

The P class of pumps is named for the phosphorylation of one of the subunits of the transport protein that occurs during the transport process. The transported substrates are moved through the phosphorylated subunit of the transport protein. A member of this class is the **sodium-potassium ATPase** that is present on the plasma membranes of all animal cells (Figure 14.3). It functions to maintain extracellular sodium concentrations and intracellular potassium concentrations. Three sodium ions are pumped out of the cell and two potassium ions are pumped in for each ATP hydrolyzed by the sodium-potassium ATPase.

Other members of the P class are calcium ATPases that pump calcium out of the cytosol into either the extracellular environment or the intracellular storage locations. Even very small increases in the concentration of free calcium ions within the cytosol can cause cellular responses to occur. Maintaining the concentration of free calcium ions in the cytosol at a low level is an important function of these calcium ATPases.

> ## Clinical Application 14.1: Inhibition of the Sodium-Potassium ATPase to Decrease Heart Rate
>
> Cardiac glycosides are inhibitors of the sodium-potassium ATPase and include ouabain and digoxin. These agents prevent cells from maintaining their normal sodium-potassium balance. When heart cells (myocytes) are exposed to a cardiac glycoside, there is an increased intracellular concentration of sodium since sodium is not pumped out by the inhibited sodium-potassium ATPase. The sodium ion gradient is much less than normal. Thus, transport via a sodium-calcium exchanger is altered since it depends on the sodium gradient in order to transport calcium out of cells. The intracellular concentration of calcium then rises since less calcium is transported out of the cell. As a result, there is an increased cardiac action potential (the electrical signal generated by nerves that results from changes in the permeability of nerve cells to certain ions), an increase in the force of contraction, and a decrease in heart rate. Drugs such as digoxin are now most commonly used to treat atrial fibrillation, an abnormal heart rhythm involving the upper chambers (atria) of the heart.

B. F-class and V-class pumps

Both F-class and V-class pumps transport **protons** (H^+). V-class pumps maintain the low pH of lysosomes by pumping protons into the lysosomes, against their electrochemical gradient, in an ATP-dependent process. Bacterial F-class pumps transport protons. In mitochondria, the F-class pump known as **ATP synthase** works in reverse. There, movement of electrons between protein complexes allows protons to be pumped from the mitochondrial matrix to the intermembrane space, creating an electrical gradient across the inner mitochondrial membrane (with more positive charges on the outside of the membrane) and also a pH gradient (the outside of the membrane is at a lower pH than the inside). The proton gradient is used by ATP synthase to drive synthesis of ATP from ADP + P_i by allowing for passive flux of protons across the membrane back into the matrix along their concentration gradient (see also *LIR Biochemistry*, pp. 77–80).

C. ABC-class pumps

The fourth class of primary active transport proteins is the ATP-binding cassette (ABC) superfamily. This name is derived from ABCs characteristic of these proteins. All ABC proteins have two cytosolic ATP-binding domains and two transmembrane domains that form the passageway for the transported molecules (Figure 14.4). The ABC cassettes bind and hydrolyze ATP, causing conformational changes in the membrane-spanning domains, inducing the translocation of substrate from one side of the membrane to the other. Seven families of ABC transporters have been characterized. All ABC transporters are involved in transport of ions, drugs, or xenobiotic compounds (natural substances that are foreign to the human body). CFTR, the cystic fibrosis transmembrane regulator, is a chloride ion channel defective in cystic fibrosis, which is a unique ABC transporter since it functions as an ion channel. CFTR uses ATP to regulate the flux of chloride ions. The molecular basis for CFTR's activity as an ATPase continues to be explored (see also Chapter 13 for more discussion of CFTR).

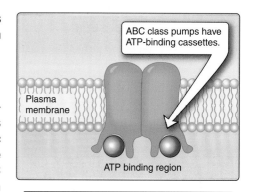

Figure 14.4
ABC-class transporters have ATP-binding cassettes.

Clinical Application 14.2: ABC Transporters and Multidrug Resistance

Cells exposed to toxic compounds, including some drugs, can develop resistance to them by decreasing their uptake, by increasing their detoxification, by altering target proteins of the toxin, and/or by increasing excretion of the drug. In this way, cells can become resistant to several drugs in addition to the initial compound. Cells resistant to a drug no longer respond to its therapeutic effects. This phenomenon is known as multidrug resistance (MDR) and is a major limitation to cancer chemotherapy. Cancer cells often become resistant to the effects of several different drugs designed to kill them. The ABC transporter ABCB1 or P-glycoprotein is associated with MDR. Inhibitors of ABCB1 have been developed, and clinical research to attempt to block the development of drug resistance has been conducted. The high concentrations of inhibitors necessary to inhibit ABCB1 are often toxic to the patient. Investigation of new approaches to circumvent MDR continues to be explored.

III. SECONDARY ACTIVE TRANSPORT

In addition to ATP-powered pumps, cells may use an active form of transport powered by the energy stored in electrochemical gradients. Ion concentration gradients of protons and sodium, generated by primary active transport, can power the movement of substrates against their gradients (Figure 14.5). Because the transport of one solute is coupled to and dependent upon the transport of another solute, this process is

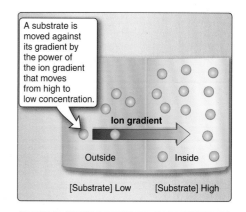

Figure 14.5
Cotransport of substrates in secondary active transport.

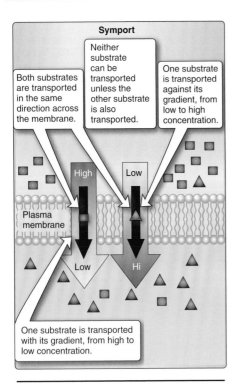

Figure 14.6
Symport involves cotransport of substrates in the same direction across membranes.

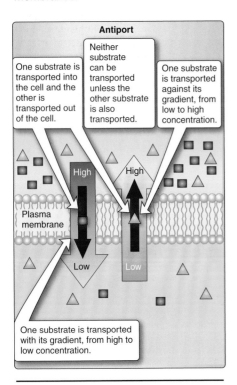

Figure 14.7
Antiport involves cotransport of substrates in opposite directions across membranes.

described as **cotransport**. Transport proteins that function in secondary transport do not possess ATPase activity. Instead, they depend indirectly on ATP hydrolysis since it is required for the primary active transport that establishes the ion gradients that power secondary active transport. Cotransport can allow substrates to cross the membrane in the same direction as each other or can allow for entry of one substrate and exit of another. While **uniporters** transport one type of molecule by facilitated transport, **symporters** and **antiporters** are cotransporters that function in secondary active transport.

A. Symporters

Symporters are secondary active transporters that move substrates in the same direction across plasma membranes (Figure 14.6). Both substrates will enter or both substrates will exit a cell by symport. One substrate is transported in an energetically favorable direction along its concentration gradient established by primary active transport. The second substrate is actively transported, against its gradient, with the energy of the concentration gradient of its cotransported substrate powering the process. The sodium-glucose transporter is a well-described symporter that transports glucose against its concentration gradient into intestinal epithelial cells using the strong concentration gradient of sodium to power the transport. Both glucose and sodium are taken into the cells (see also Chapter 15 for more discussion of glucose transport). Sodium-dependent amino acid transport proteins also function in symport.

B. Antiporters

Antiporters cotransport molecules in opposite directions across plasma membranes (Figure 14.7). The movement of one substrate into a cell is coupled to the movement of another substrate out of the cell. One substrate is moved along its gradient and the other moved against its gradient. The sodium-calcium antiporter in cardiac muscle cells is an example of an antiporter. This system maintains low calcium in the cytosol so that an increase in calcium can then trigger muscle contraction. Three sodium ions are moved into the cell, along the sodium gradient, while one calcium ion is moved out of the cell, against its gradient. The sodium-calcium antiporter thereby decreases the cytosolic concentration of calcium and reduces the strength of heart muscle contraction (see also information about inhibition of the sodium-potassium ATPase earlier in this chapter). Another antiporter is the sodium-proton exchanger that catalyzes the electroneutral exchange of sodium ions and protons and regulates salt concentration and pH. The bicarbonate/chloride antiporter in the stomach is another example.

Chapter Summary

- Active transport involves the movement of ions or molecules against their concentration gradient in an energy-dependent process.
- Primary active transport requires the direct hydrolysis of ATP by the primary active transport protein.
- Four classes of ATPases function in primary active transport. P-class transporters move ions against their gradients, F-class and V-class transporters pump protons against their concentration gradients, and ABC-class transporters move drugs, ions, and xenobiotics against their gradients.
- Ion gradients are generated and maintained by primary active transport.
- The ion gradients established by primary active transport can drive the transport of ions and small molecules against their concentration gradients in secondary active transport.
- Symporters are secondary active transporters that move substrates in the same direction (to the inside or to the outside) across membranes.
- Antiporters move one substrate into the cell and another substrate out of the cell by secondary active transport.

Study Questions

Choose the ONE best answer.

14.1 A membrane-spanning protein hydrolyzes ATP when it transports its substrate across the plasma membrane of the cell against its concentration gradient. Which of the following is the most likely identity of this membrane protein?

A. ABC transporter
B. Bicarbonate/chloride antiporter
C. Glucose uniporter
D. Sodium-dependent amino acid transporter
E. Sodium-glucose transporter

Correct answer = A. The protein described is an ATPase pump that functions in primary active transport. ABC transporters are ATPase pumps that transport drugs and function in glucose transport. GLUT5 transports fructose and not glucose. Both SGLT1 and SGLT2 are sodium-dependent glucose transporters that transport glucose against its concentration gradient. The bicarbonate/chloride transporter is an antiporter that functions in secondary active transport and does not directly hydrolyze ATP. A glucose uniporter transports glucose by facilitated diffusion, without ATP hydrolysis and along the concentration gradient of glucose. The sodium-glucose and sodium–amino acid transporters are both symporters that use secondary active transport and do not hydrolyze ATP themselves.

14.2 In the membrane transport system that involves movement of 3 sodium ions out of cells against their concentration gradient and 2 potassium ions into cells against their concentration gradient

A. ATP is hydrolyzed by the transporter to power movement of the ions.
B. Glucose is also transported into cells by the same transport protein.
C. Insulin signaling causes the transporter to move to the cell's surface.
D. Ion gradients created by secondary active transport power the system.
E. Simple diffusion is used to transport the ions into and out of the cells.

Correct answer = A. This transport system for sodium and potassium utilizes an ATP-powered pump to transport the ions against their gradients. Glucose is not also transported and insulin is not involved. This is a primary active transport system. Ion gradients that are established by primary active transport sometimes drive secondary active transport, but the reverse does not occur. Membrane transport of sodium and potassium into individual cells does not occur via simple diffusion.

14.3 The sodium-potassium ATPase is inhibited by a drug. Which of the following consequences may be expected in response to this inhibition?

   A. Buildup of potassium within the cells
   B. Excess loss of sodium from the cells
   C. Failure to establish an adequate sodium gradient
   D. Inability to pump protons into the cell
   E. MDR of the sodium-potassium ATPase

Correct answer = C. Inhibition of the sodium-potassium ATPase would impair the establishment of sodium gradients normally generated and maintained by this primary active transporter. Sodium ions are pumped out of cells by the sodium-potassium ATPase, while potassium ions are normally pumped into cells. Inhibition of the ATPase would result in less sodium being pumped outside of cells and less potassium being pumped inside cells. The sodium-potassium ATPase does not pump protons. MDR can develop when members of the ABC class of ATPases pump out drugs that are being delivered as therapy, causing the cells to become resistant. The sodium-potassium ATPase is a P-class ATPase and not an ABC transporter. Inhibition of the sodium-potassium ATPase would not cause the sodium-potassium ATPase to become resistant to drugs.

14.4 A 48-year-old male patient with lung cancer is given chemotherapy treatment. Initially, the treatment appears to be successful. With time, MDR develops. Which of the following consequences would be expected?

   A. Increased transport of drugs into the cancer cells
   B. Inhibition of ABC transporters on the cancer cells
   C. Reduced hydrolysis of ATP to ADP and P_i
   D. Secondary active transport of drugs out of the cells
   E. Survival of the cancer cells

Correct answer = E. Cancer cells survive and are not killed by chemotherapy drugs when MDR develops. ABC transporters move the drugs out of the cancer cells so that the drugs do not cause damaging effects to the cancer cells. ABC transporters are not inhibited, and their ability to hydrolyze ATP is not impaired. In fact, ABC transporters are more active when MDR occurs. Secondary active transport is not the mechanism by which drugs are transported out of cells in MDR. ABC transporters are used, and they hydrolyze ATP as primary active transporters.

14.5 Sodium-dependent amino acid transporters are symporters that allow for movement of amino acids into cells against their concentration gradient. The energy for this process is derived from

   A. ATP hydrolysis by the sodium-dependent amino acid transporter.
   B. A sodium ion concentration gradient.
   C. Antiport of amino acids with sodium ions.
   D. Hydrolysis of ATP by the sodium-dependent amino acid transporter.
   E. Primary active transport.

Correct answer = B. A sodium ion gradient (generated and maintained by the sodium-potassium ATPase) drives the transport of amino acids against their concentration gradient by the sodium-dependent amino acid symporter. ATP is not hydrolyzed by symport proteins as they function in secondary active transport. Amino acids are symported with sodium, not antiported. Antiport would not power the process. Symport is a form of secondary active transport and not primary active transport.

14.6 The sodium-proton exchanger is an antiporter. Therefore, it transports

   A. Both sodium ions and protons against their concentration gradients.
   B. By hydrolyzing ATP to power movement of both substrates.
   C. Individual substrates one at a time across the plasma membrane.
   D. Sodium ions and protons in opposite directions across a membrane.
   E. Sodium into the cell against its concentration gradient.

Correct answer = D. Antiporters function in secondary active transport to move substrates in opposite directions across a membrane. In antiport, one substrate will be moved along its concentration gradient. Antiporters are cotransporters that move both substrates together and not individually across membranes. ATP is not directly hydrolyzed during secondary active transport by an antiporter. The gradient for sodium is higher outside cells (owing to the sodium-potassium ATPase) than inside cells.

14.7 The energy used to transport glucose against its concentration gradient into intestinal epithelial cells results from

A. Direct hydrolysis of ATP by the sodium-glucose transporter.

B. A glucose gradient between the intestinal epithelial cell and the blood.

C. The sodium gradient established by the sodium-potassium ATPase.

D. Primary active transport using an ABC-class pump.

E. Differences in osmotic pressure on either side of the cell membrane.

Correct answer = C. The sodium gradient established by the sodium-potassium ATPase drives transport of glucose against its concentration gradient and into intestinal epithelial cells. The sodium-glucose transporter is a symport protein that functions in secondary active transport and does not have ATPase activity of its own. It is not a primary active transport protein. There is a higher concentration of glucose inside the intestinal epithelial cells than in the intestinal lumen, so uniport, which involves movement from high to low concentration, cannot be used. Differences in osmotic pressure on either side of a plasma membrane stimulate water transport via aquaporins in the process known as osmosis.

Glucose Transport 15

	Location	Function
GLUT 1	Most tissues	Basal glucose uptake
GLUT 2	Liver, kidneys, pancreas	Removes excess glucose from blood
GLUT 3	Most tissues	Basal glucose uptake
GLUT 4	Muscle and fat	Removes excess glucose from blood
GLUT 5	Small intestine, testes	Transport of fructose

Figure 15.1
Tissue distribution of glucose transporters.

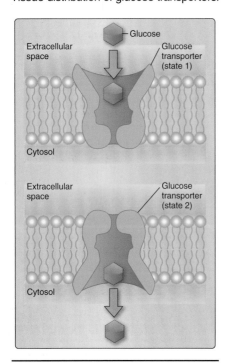

Figure 15.2
Facilitated transport of glucose through a plasma membrane.

I. OVERVIEW

Glucose is essential for life since it is the major energy source for mammalian cells. Transport of glucose into cells is crucial for survival of both the cell and the individual. Most cells use facilitated transport for the uptake of glucose by uniport, since a concentration gradient often exists for glucose between extracellular fluids and the cytoplasm of cells. A family of glucose transport proteins in the GLUT family of proteins encoded by *SLC2* genes catalyze facilitated transport of glucose. This family is divided into three classes, grouped by sequence similarity, and distinct GLUT proteins are expressed in most cell types. In other types of cells, such as intestinal epithelial cells, glucose must be transported against its concentration gradient. There, an active form of glucose transport is necessary, and another family of transport proteins, SGLT proteins, facilitate secondary active transport of glucose (see also *LIR Biochemistry*, p. 97).

II. FACILITATED TRANSPORT OF GLUCOSE

The family of glucose transporters (GLUTs) that functions in facilitated transport of glucose includes at least 14 members, 11 of which are involved in glucose transport. GLUTs 1 to 5 are the most commonly expressed. **GLUT**s have specific tissue locations with an affinity for glucose that permits their function within a particular tissue environment (Figure 15.1). GLUTs 1, 3, and 4 are primarily involved in glucose uptake from the blood. GLUT1 and GLUT3 are found in most tissues. GLUT4 is found in skeletal muscle and adipose (fat) cells. GLUT2 transports glucose into liver and kidney cells when blood glucose levels are high and transports glucose out of cells into the blood when blood glucose levels are low. GLUT2 is also found in pancreatic β cells. GLUT5 is the primary transporter for fructose (instead of glucose) in the small intestine and testes.

A. Mechanism of glucose transport

GLUT family members transport glucose from an area of higher concentration of glucose to an area of lower concentration of glucose. In order to be transported, the glucose molecule binds to the GLUT protein (state 1), and then the membrane direction of the glucose-GLUT complex is reversed (state 2) so that the sugar is released to the other side of the membrane (Figure 15.2). After the sugar has dissociated, the GLUT flips orientation and is readied for another cycle of transport.

B. General considerations

Glucose transport depends upon the **concentration of glucose** and on the **number of GLUT proteins** present on the plasma membrane. GLUT family members have the capacity to transport glucose in both directions across membranes. However, in most cells, the cytoplasmic glucose concentration is lower than the extracellular concentration of glucose. Transport of glucose can proceed only along (down or with) its concentration gradient, from the exterior environment into the cell. In liver and kidney cells that synthesize glucose (via gluconeogenesis), the intracellular concentration of glucose can be greater than the blood glucose concentration surrounding the cell (see also *LIR Biochemistry,* Chapter 10, for discussion of gluconeogenesis). Then, transport of glucose out of liver and kidney cells occurs. GLUT2 proteins export glucose from these cell types.

C. Role of insulin

Insulin is a regulatory hormone secreted by β cells of the islets of Langerhans in the pancreas in response to glucose (and other carbohydrates) and amino acids (Figure 15.3). Epinephrine, a hormone released in response to stress, inhibits insulin release (see also *LIR Biochemistry,* Chapter 23). Insulin coordinates the use of fuel by tissues and is required for normal metabolism. The importance of insulin to health is illustrated by **type 1 diabetes mellitus** in which β cells of affected individuals are destroyed in an autoimmune response, halting endogenous insulin production (Figure 15.4). Individuals with type 1 diabetes mellitus must receive exogenous insulin. Insulin's metabolic effects are **anabolic**, favoring the building of stores of carbohydrates, fats, and proteins. Failure of cells to respond properly to insulin, known as insulin resistance, is characteristic of **type 2 diabetes mellitus**. Insulin therapy may or may not be required by individuals with type 2 diabetes. Therapies for type 2 diabetes include agents designed to decrease blood glucose (hypoglycemic agents) as well as diet and exercise. (Weight loss often improves insulin resistance.) One important normal response to insulin is to initiate transport of glucose into certain cell types.

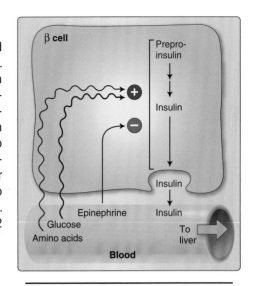

Figure 15.3
Regulation of insulin release from pancreatic β cells.

	Defect	Treatment
TYPE 1	β cells destroyed, no insulin produced	Insulin
TYPE 2	Insulin resistance	Diet, exercise, oral hypoglycemic agents Insulin may or may not be necessary

Figure 15.4
Comparison of type 1 and type 2 diabetes mellitus.

Clinical Application 15.1: Glucose Transport and Insulin Secretion

GLUT2 glucose transporters are found in pancreatic β cells that must sense and respond to a higher blood concentration of glucose by secreting insulin. GLUT2 transporters are low-affinity glucose transporters (with high K_m) and transport glucose only when the concentration of circulating glucose is greater than 5 mM, the normal concentration of glucose in blood. Glucose is then transported into the β cells when glucose levels in the blood increase from basal levels after consumption of carbohydrates. The transport of glucose into the β cells causes an increase in their intracellular glucose concentration, promoting release of insulin within storage vesicles within the β cells.

1. **Insulin-insensitive glucose transport via GLUT1, GLUT2, and GLUT3:** Most cell types do not require insulin in order to transport glucose down its concentration gradient into cells (Figure 15.5).

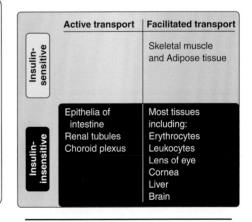

	Active transport	Facilitated transport
Insulin-sensitive		Skeletal muscle and Adipose tissue
Insulin-insensitive	Epithelia of intestine Renal tubules Choroid plexus	Most tissues including: Erythrocytes Leukocytes Lens of eye Cornea Liver Brain

Figure 15.5
Characteristics of glucose transport in various tissues.

These cell types include erythrocytes, leukocytes, and liver and brain cells, which express insulin-insensitive GLUT1, GLUT2, and/ or GLUT3 receptors on their cell surfaces at all times.

2. **Insulin-sensitive glucose transport via GLUT4:** In skeletal muscle and fat cells (adipocytes), GLUT4 proteins normally reside inside cells, within intracellular vesicles, inactive and bound to the Golgi complex (Figure 15.6). When a resting skeletal muscle cell or adipocyte is stimulated by insulin, the vesicles containing GLUT4 translocate to the surface to function in glucose transport. In exercising skeletal muscle cells, muscle contraction induces GLUT4 translocation to the membrane from exercise-sensitive vesicles.

Expression of GLUT4 on the membrane surface only at certain times plays a rate-limiting role in glucose utilization by skeletal muscle and adipose cells where glucose transport is stimulated by a series of events. Consumption of carbohydrates stimulates insulin release from β cells of the pancreas. Insulin circulates throughout the blood and binds to insulin receptors on many cell types. When the insulin receptors on resting skeletal muscle cells and adipocytes are bound by insulin, then intracellular signaling pathways are initiated by insulin and stimulate movement of GLUT4 from intracellular vesicles to the membrane surface. A series of highly organized trafficking events is involved, including actin remodeling beneath the plasma membrane so that GLUT4-containing vesicles can fuse with the plasma membrane. Once GLUT4 is expressed on the surface of the cell, then glucose transport occurs as long as a glucose concentration gradient exists. GLUT4 proteins are removed from the plasma membrane by endocytosis and recycled back to their intracellular storage compartment when insulin stimulation is withdrawn.

Clinical Application 15.2: Glucose Transport in Diabetes Mellitus

Individuals with type 1 diabetes mellitus do not produce insulin. Approximately 10% of individuals with diabetes mellitus in the United States have the type 1 form. Their pancreatic β cells have been destroyed as a consequence of autoimmune reactions, and insulin production is no longer possible. Insulin must be supplied exogenously to control hyperglycemia (high blood levels of glucose) and to prevent diabetic ketoacidosis (breakdown of fats that occurs in the absence of insulin with the potentially life-threatening consequence of acidifying the blood). In order for glucose transport to occur into the insulin-dependent skeletal muscle and fat cells, exogenous insulin must be administered. Otherwise, glucose cannot enter those cells, and elevated concentrations of glucose are found in the blood.

In persons with type 2 diabetes mellitus, peripheral tissues have become resistant to the effects of insulin. Insulin secretion may be impaired, but persons with type 2 diabetes mellitus do not require insulin to sustain life. In this group that constitutes approximately 90% of persons with diabetes mellitus in the United States, insulin signaling via insulin receptors is ineffective. In skeletal muscle and adipose, insulin-sensitive glucose transport will be impaired. In the liver, cells will not respond to insulin by inhibiting gluconeogenesis (endogenous production of glucose). While insulin is present, often in high concentrations, the cells of individuals with type 2 diabetes mellitus do not respond to it. Drug therapies for type 2 diabetes include agents that improve insulin signaling (see also *LIR Biochemistry*, Chapter 25, for a discussion of diabetes mellitus and Chapter 18 for a discussion of insulin signaling). Long-standing elevation of blood glucose in both types of diabetes mellitus causes chronic complications including premature atherosclerosis (including cardiovascular disease and stroke), retinopathy, nephropathy, and neuropathy.

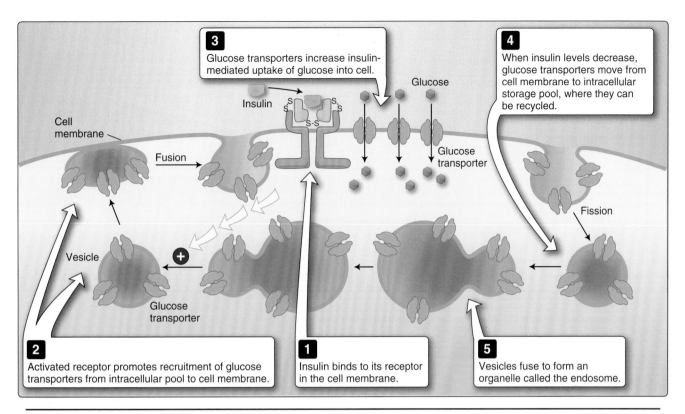

Figure 15.6
Insulin causes some cells to recruit transporters from intracellular stores.

III. ACTIVE TRANSPORT OF GLUCOSE

There are three main locations where glucose is found at a higher concentration inside cells than out and glucose must be transported against its gradient to enter those cells. The first is the **choroid plexus**, the second is the **proximal convoluted tubules of the kidney**, and the third is the **epithelial cell brush border of the small intestine**. The epithelial cells lining the small intestine are a barrier between the lumen of the small intestine and the bloodstream (Figure 15.7). Dietary glucose must be absorbed by the epithelial cells and then transferred to the underlying connective tissue so that it can enter into the blood circulation. GLUT proteins cannot be used to transport glucose into these cells because glucose is at a higher concentration inside the cells than outside the cells. (Note that the volume of the small intestine is much greater than the volume of an epithelial cell and that concentration is dependent on volume.) The apical membranes (facing the lumen) of the epithelial cells contain **sodium-glucose transport proteins (SGLT)** that catalyze transport of glucose against its concentration gradient. Six members of the SGLT family have been reported but only SGLT1 and SGLT2 are well characterized.

This process of glucose transport against its concentration gradient is an example of **secondary active transport** (see also Chapter 13). Glucose transport by SGLTs can only occur when sodium is present

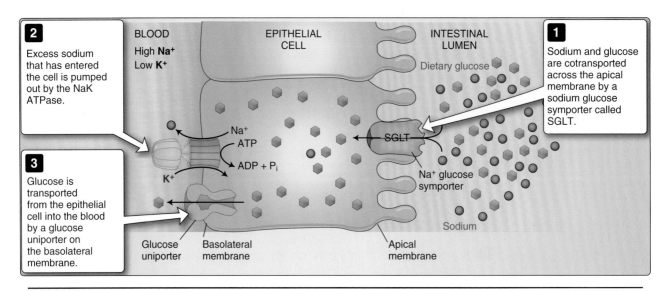

2

Excess sodium that has entered the cell is pumped out by the NaK ATPase.

3

Glucose is transported from the epithelial cell into the blood by a glucose uniporter on the basolateral membrane.

1

Sodium and glucose are cotransported across the apical membrane by a sodium glucose symporter called SGLT.

Figure 15.7
Glucose transport from the intestinal lumen to the bloodstream.

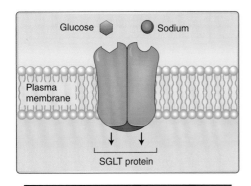

Figure 15.8
Sodium-glucose transporters cotransport sodium and glucose in a symport process.

and is cotransported with glucose (Figure 15.8). Since both glucose and sodium are transported in the same direction (both into the cell), this is a **symport** process. The energy required to power the transport of glucose against its gradient is derived from the electrochemical gradient of sodium that is created and maintained by the sodium-potassium ATPase, a primary active transport system (see also Chapter 14). Since sodium has such a strong concentration gradient from outside to inside the cell, its tendency to move down its gradient is great. On the apical membrane of intestinal epithelial cells, the only transporter for sodium is a sodium-glucose transporter. Sodium is transported only when glucose is cotransported along with it. Glucose effectively gains a free ride into the cell owing to the strong concentration gradient of sodium. Once inside the cell, the glucose can then be transported out of the cell and into the blood by a GLUT protein on the basolateral (opposite to apical) membrane.

Clinical Application 15.3: Oral Rehydration Therapy Based on Glucose-Sodium Cotransport

In 1960, an American biochemistry named Robert Crane first described glucose-sodium transport as a mechanism of intestinal absorption of glucose. This discovery was the foundation for development of a therapy known as oral rehydration. A glucose-containing salt solution administered to patients who are dehydrated by effects of cholera is effective therapy. It allows the glucose to accelerate the absorption of solute and water and counterbalances the loss of water and electrolytes caused by cholera toxin. Oral rehydration therapy is credited with saving the lives of millions of patients with cholera in developing nations since the 1980s.

Clinical Application 15.4: Glucose-Sodium Cotransport in the Kidney, Implications in Diabetes Treatment

Under normal circumstances, the kidney filters glucose in the glomerulus and into the Bowman space and the renal tubules. The amount filtered is related to the glucose concentration in blood. All this glucose is normally reabsorbed in the proximal tubule (by transport into the epithelial cells lining the proximal tubule) with none appearing in urine. Blood glucose concentration is normally approximately 5 mM, and epithelial cells within the proximal tubule have a glucose concentration of 0.05 mM. The glucose must cross through these epithelial cells into the interstitial fluid (5 mM glucose) and back into blood. The gradient up which glucose must be pumped is 0.05 to 5 mM. As this process continues, more and more glucose is removed from the tubular fluid, and the concentration drops as low as 0.005 mM, increasing the concentration gradient tenfold. Secondary active transport is used to move glucose up this gradient. A sodium-glucose symport protein, SGLT2, is used. SGLT2 is responsible for 90% of the glucose transported (reabsorbed) in the first segment of the proximal tubules, with one sodium ion transported per glucose molecule. SGLT1 transports the remaining 10% of glucose in the distal segments of the proximal tubules, transporting 2 sodium ions with every glucose molecule. The sodium gradient is generated by the sodium-potassium ATPase pump; the concentration of sodium outside the cells is approximately 140 mM, and the concentration inside is approximately 10 mM.

In individuals with diabetes, elevated blood glucose levels cause a greater amount of glucose to be filtered by the kidney. The ability of the kidney to reabsorb all of the glucose is often exceeded and glucose spills into the urine. The amount of glucose appearing in the urine is the amount that exceeds the reabsorption capacity of the kidney. Inhibition of SGLT2 in individuals with type 2 diabetes leads to decreased reabsorption of glucose and increased excretion of glucose into urine. Increased glucose excretion in urine results in decreased plasma glucose levels and excretion of calories (from glucose) into the urine, with the potential for weight loss. Weight loss often improves the insulin resistance seen in type 2 diabetes. Several SGLT2 inhibitors are now available to treat individuals with type 2 diabetes mellitus. This class of medicines, approved by the Food and Drug Administration in the United States, is used along with diet and exercise to lower blood glucose in adults with type 2 diabetes.

Chapter Summary

- Most cells use facilitated transport of glucose mediated by GLUT proteins.
- Glucose is moved from higher concentration (usually outside the cell) toward an area of lower concentration (usually inside the cell).
- Most cells have insulin-independent glucose transport.
- Skeletal muscle and adipose cells require insulin to stimulate movement of GLUT4 proteins from intracellular vesicles to become inserted in the plasma membrane where they can function to take up glucose.
- If insulin is absent (type 1 diabetes) or does not signal properly (type 2 diabetes), then insulin-dependent glucose transport by GLUT4 will cease or be impaired.
- Secondary active transport of glucose occurs via cotransport with sodium, using SGLT proteins, in the choroid plexus, proximal tubules of kidneys, and the intestine. Since sodium and glucose are moved in the same direction across the plasma membrane, this is a symport process.

Study Questions

Choose the ONE best answer.

15.1 Glucose transport proteins are observed on a cell's surface. The intracellular glucose concentration is 2 mM, and the extracellular concentration is raised to 5 mM. The transporters begin to transport glucose from outside to inside the cell. Which of the following is the type of glucose transport protein involved in this transport?

A. GLUT1
B. GLUT4
C. GLUT5
D. SGLT1
E. SGLT2

Correct answer = A. GLUT1 transporters reside on the cell's surface and transport glucose from an area of higher to lower concentration. GLUT4 transporters are insulin-dependent transporters in skeletal muscle and liver. They do not reside on the cell's surface prior to their function in glucose transport. GLUT5 transports fructose and not glucose. Both SGLT1 and SGLT2 are sodium-dependent glucose transporters that transport glucose against its concentration gradient.

15.2 Glucose transport from the blood into hepatocytes (liver cells) requires

A. ATP hydrolysis by a symporter that creates a sodium ion gradient.
B. Function of a primary active transporter to create a glucose gradient.
C. Higher blood concentration of glucose than hepatocyte concentration of glucose.
D. Insulin stimulation of GLUT translocation to the hepatocyte plasma membrane.
E. Uniport of glucose against its concentration gradient from blood into hepatocytes.

Correct answer = C. Glucose transport into hepatocytes occurs via uniport and therefore requires a higher blood concentration of glucose than hepatocyte concentration of glucose. The glucose is then transported along/down/with its gradient from high to lower concentration of glucose. Glucose transport does not occur via primary active transport. ATP hydrolysis is not required for glucose transport via uniport. And uniport does not occur against the concentration gradient. GLUTs are present constitutively on the surface of hepatocytes and do not require insulin stimulation for their translocation to the cell surface.

15.3 A hepatocyte increases its transport of glucose through GLUT2 transporters when blood glucose levels increase to 20 mM. However, glucose transport through GLUT1 in an erythrocyte continues at the same rate as it did at lower blood glucose concentrations. Which of the following best accounts for these findings?

A. GLUT1 has a larger K_m than does GLUT2.
B. GLUT1 glucose transport requires cotransport with sodium.
C. GLUT2 achieves its V_{max} at a glucose concentration less than 20 mM.
D. GLUT2 glucose transport requires insulin activation.
E. GLUT2 has a lower glucose affinity than does GLUT1.

Correct answer = E. Based on this information, we can determine that GLUT2 has a lower glucose affinity than does GLUT1. Lower affinity transporters have higher K_m values. GLUT1 is already at V_{max} at a glucose concentration less than 20 mM. These are higher affinity, lower K_m glucose transporters than GLUT2. In order for GLUT2 to be able to increase its rate of transport when glucose concentration is increased, its K_m must be greater than the usual blood glucose concentration. Insulin is not needed, and cotransport of glucose with sodium does not occur in hepatocytes.

15.4 A resting skeletal muscle cell is carrying out insulin-dependent glucose transport. Insulin's role in this process is to promote

A. Cotransport of glucose with sodium against the glucose concentration gradient.
B. Hydrolysis of ATP to power glucose transport against its concentration gradient.
C. Movement of glucose from inside the cell to the extracellular environment.
D. Primary active transport of glucose with potassium into the cell.
E. Translocation of GLUT4 proteins from intracellular vesicles to the cell's surface.

Correct answer = E. In resting skeletal muscle cells (and in adipocytes), insulin promotes movement of GLUT4 transporters from intracellular vesicles to the surface of the cell to allow for glucose transport from high to low concentration of glucose. Insulin is not involved with cotransport of glucose with sodium. Insulin does not promote ATP hydrolysis or movement of glucose from inside to outside of cells. Glucose is not transported by primary active transport with potassium.

15.5 A 56-year-old male with a history of type 2 diabetes mellitus has a fasting blood glucose of 150 mg/dL (reference range 70 to 100 mg/dL). In this individual, transport of glucose is impaired into which of the following cell types?

A. Adipocytes
B. Brain cells
C. Eye cells
D. Liver cells
E. Red blood cells

Correct answer = A. Adipocytes (fat cells) and resting skeletal muscle cells have insulin-dependent glucose transport. If insulin signals are not received properly, as in type 2 diabetes, insulin-dependent glucose transport into adipocytes and also into resting skeletal muscle cells will be impaired. Brain cells, eye cells, liver cells, and red blood cells have insulin-independent glucose transport.

15.6 The energy to power transport glucose against its concentration gradient into intestinal epithelial cells is derived from

A. ATP hydrolysis by the SGLT.
B. A sodium ion concentration gradient.
C. Antiport with sodium ions.
D. GLUT hydrolysis of ATP.
E. Primary active transport.

Correct answer = B. A sodium ion gradient (generated and maintained by the sodium-potassium ATPase) drives the transport of glucose against its concentration gradient into the intestinal epithelial cells. Neither SGLTs nor GLUTs hydrolyze ATP. Glucose is transported by symport with sodium, not by antiport with sodium. Glucose-sodium symport is a form of secondary active transport and not primary active transport since ATP is not directly hydrolyzed by the transport protein itself.

15.7 A drug designed to inhibit SGLT2 in the kidneys of individuals with type 2 diabetes functions in part by increasing

A. Reabsorption of glucose.
B. Glucose secretion into urine.
C. Glucose transport into epithelial cells.
D. Primary active transport of glucose.
E. Sodium transport into epithelial cells.

Correct answer = B. Inhibition of the SGLT sodium-glucose symporter in kidneys results in more glucose being secreted into urine since absorption of glucose is impaired. Glucose absorption is not increased but decreased because transport of glucose into the epithelial cells is being inhibited. Sodium transport is also inhibited and not increased. Inhibition of SGLT results in an increased function of any primary active transporter. Glucose is never transported by primary active transport. SGLTs facilitate secondary active transport of glucose.

Drug Transport

16

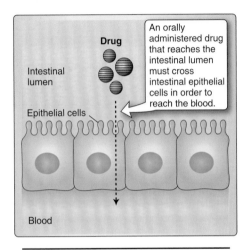

Figure 16.1
Orally administered drugs must penetrate the epithelial cells of the intestinal mucosa in order to enter the circulation.

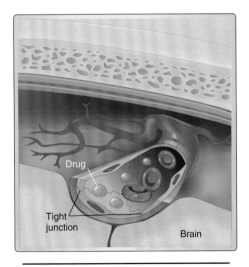

Figure 16.2
The blood-brain barrier restricts entry of drugs to the central nervous system.

I. OVERVIEW

When a drug is administered intravenously, its full dose reaches the systemic circulation. Since drugs must enter the bloodstream in order to reach cells that are targets of their actions, intravenous administration ensures this quick entry into the blood. But drugs administered in other ways may be only partially absorbed into the bloodstream. Oral administration is the most common route and requires that a drug first dissolve in the gastrointestinal fluid and then penetrate the epithelial cells of the intestinal mucosa in order to access the bloodstream (Figure 16.1). Drugs designed to reach the central nervous system must additionally cross the blood-brain barrier formed by endothelial cells that line blood vessels in the central nervous system. These cells form tight junctions that restrict the entry of molecules greater than 400 daltons in size. This barrier is a major obstacle for development of drugs to treat central nervous system disorders including neurodegenerative diseases (such as multiple sclerosis, Alzheimer disease, and Parkinson disease); psychiatric illnesses (such as anxiety, depression, and schizophrenia); and stroke and cerebrovascular disorders (Figure 16.2).

Most drugs are either weak acids or weak bases. Concentrations of permeable forms of drugs are sometimes determined by relative concentrations of their charged and uncharged forms since uncharged molecules are thought to pass through membranes more readily than charged molecules. Studies of relationships of acid and base concentrations to pH have been done to determine the amount of drug that will be found on each side of a membrane (see also *LIR Pharmacology*, pp. 6–8). While it has long been assumed that lipid-soluble drugs can diffuse across cell membranes, barriers to diffusion do exist in the hydrophilic head groups of membrane phospholipids (see also Chapter 13). Some small water-soluble molecules may be able to use aquaporins (water channels) to cross membranes (Figure 16.3). It is increasingly recognized that transport proteins facilitate the movement of drugs across biological membranes. Active transport processes appear to be most commonly used by drugs.

II. CLASSES OF DRUG TRANSPORTERS

Many drug transporters function as primary and secondary **active transporters** (see Chapter 15 for a discussion of active transport). Based on sequence similarities, drug transporters have been classified as **solute carriers (SLCs)** and **ATP-binding cassette (ABC) transporters** (Figure 16.4). The structures, functions, and tissue distributions of drug transporters within each group can vary widely.

A. Solute carriers

The SLCs are classified into four major families: peptide transporters (PEPT), organic anion-transporting polypeptides (OATP), organic ion transporters, and H+/organic cation antiporters.

1. **Peptide transporters:** Protons (H+) and peptides are cotransported across membranes by PEPT proteins that transport small peptides (two and three amino acids) but not individual amino acids or larger peptides. PEPT proteins transport drugs such as β-lactam antibiotics (including penicillin) and ACE inhibitors (**a**ngiotensin-**c**onverting **e**nzyme inhibitors, used to treat hypertension) across the intestinal epithelium. In order to increase absorption of certain anticancer and antiviral drugs, their structures are being altered, with amino acid sequences that convert them into better PEPT substrates.

2. **Organic anion-transporting polypeptides:** Transporters in this group catalyze the movement of amphipathic organic compounds such as bile salts, steroids, and thyroid hormones. One member of this family, OATP1B1, is responsible for hepatic (liver) uptake of pravastatin, an inhibitor of cholesterol synthesis, and of the ACE inhibitor enalapril. Individual genetic variations in OATP1B1, known as gene polymorphisms, cause altered transport of pravastatin and differences in individual patients' responses to the drugs.

3. **Organic ion transporters:** This family makes up a large group of drug transporters. Some are uniporters, while others are symporters or antiporters. Various members of this family are expressed in the liver, kidney, skeletal muscles, and intestinal brush border. Renal secretion of drugs and toxins is mediated in part by organic ion transporters. The drug metformin (a hypoglycemic agent used to treat elevated blood glucose in type 2 diabetes) is taken up by a transporter in this family.

4. **H+/organic cation antiporters:** Organic cations are excreted by the proton/organic cation antiporter in brush border membranes. An oppositely directed proton, H+, gradient is the driving force of the transport.

B. ABC transporters

ATP-binding cassette (ABC) transporters use primary active transport to export ions and xenobiotics from cells (see also Chapter 15). These represent a main route for cells to expel toxins.

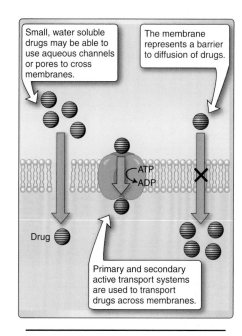

Figure 16.3
Mechanisms of drug crossing the intestinal epithelial cells.

Drug Transporters	Transport
Solute carriers (SLC's):	*Secondary active transport*
PEPT Peptide transporters	Proton (H+) and di-and tri-peptide cotransport
OATP Organic anion transporting polypeptides	Amphipathic organic compounds
Organic ion transporters	Organic ions
H+/organic cation antiporters	Organic cations are excreted and H+ taken up
ABC transporters	*Primary active transport*
	Export of ions, drugs and xenobiotics

Figure 16.4
Drug transporters include SLCs and ABC transporters.

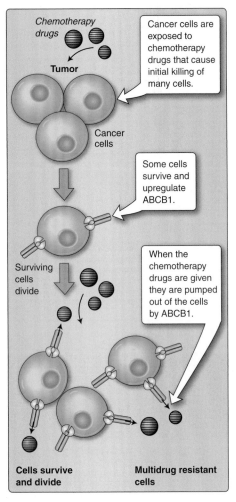

Figure 16.5
Development of MDR.

But ABC transporters can also expel drugs from cells. There is a growing list of substrates of ABC transporters including anticancer agents, antiviral agents, calcium channel blockers, and immunosuppressive agents. **Multidrug resistance** (MDR) can develop when cells expel the therapeutic agents designed to inhibit or kill them (Figure 16.5). Cancer cells that develop MDR do not respond to chemotherapy drugs. Inhibitors of ABC transporters are being explored as a way to prevent MDR. **ABCB1** or **P-glycoprotein** (P-gp) is an ABC transporter often implicated in MDR and is responsible for extruding chemotherapeutic agents from cancer cells. ABCB1 is also expressed in normal tissues. For example, in the intestinal epithelial brush border, ABCB1 is responsible for the efflux of xenobiotics before they reach the circulation. Herbals such as St. John's wort and the antituberculosis drug, rifampin, induce intestinal expression of ABCB1. Such increased expression of ABCB1 decreases the absorption of ABCB1 substrates such as the drug, digoxin (used to increase heart muscle contraction and to decrease heart rate), because, when ABCB1 is expressed at higher levels, the drugs are exported by intestinal epithelial cells.

Chapter Summary

- Drugs administered orally must cross barriers of the plasma membranes in the intestinal epithelial cells in order to be absorbed into the bloodstream.
- Endothelial cells lining blood vessels in the central nervous system form a blood-brain barrier to drugs aimed at the central nervous system.
- While passive diffusion has long been described as a mechanism of drug transport into cells, the plasma membranes of these cells actually represent a barrier to free diffusion.
- Drug transport proteins are being identified on many cell types in the body and include SLCs and ABC transporters.
- Many drug transporters use active transport, either primary or secondary.
- While SLCs transport drugs into cells, ABC transporters export drugs and xenobiotic compounds from cells.
- ABC transporters mediate MDR that develops when cells expel drugs designed to inhibit or kill them, such as cancer cells exporting chemotherapy agents.

Study Questions

Choose the ONE best answer.

16.1 An individual swallows aspirin tablets to attempt to relieve his muscle soreness. In order for the aspirin to take effect, it must

  A. Bind to P-gp to stimulate MDR.
  B. Cross intestinal epithelial cells to enter the circulation.
  C. Stimulate ATP hydrolysis of an ABC transporter.
  D. Undergo antiport from the stomach.
  E. Use an ABC transporter to enter the liver.

Correct answer = B. Drugs taken by the oral route have to cross the barrier of epithelial cell membranes in the intestine in order to enter the circulation. P-gp is an ABC transporter involved in MDR. It transports drugs out of cells and would not aid in the absorption of aspirin. ABC transporters transport drugs out of cells and not into them. Antiport from the stomach would not enable the drug to enter into the circulation since it must still cross the barrier of intestinal epithelial cells in order to enter the circulation.

16.2 A new protein drug is being developed with the goal of treating Huntington disease, a neurodegenerative disorder that impacts the central nervous system. The main obstacle to delivery of this drug to the target sites will be

  A. Absorption by the gastrointestinal tract.
  B. Endothelial cells lining blood vessels within the central nervous system.
  C. Lack of an ABC transporter to deliver the drug to appropriate cells.
  D. MDR developing quickly.
  E. Poor solubility in aqueous solutions of the body.

Correct answer = B. The main barrier to delivery of drugs to the central nervous system is the blood-brain barrier. Endothelial cells lining the blood vessels within the central nervous system form this barrier that excludes most of the large molecules (>400 daltons) from entering. Absorption by the gastrointestinal tract may occur, but if the drug cannot access the central nervous system owing to the blood-brain barrier, then effects will not be observed. ABC transporters transport drugs out of cells. Lack of expression of an ABC transporter could improve the delivery of the drug across the blood-brain barrier. MDR develops in response to actions of ABC transporters, particularly P-gp. Solubility in aqueous solutions could be very high, but if the drug cannot cross the blood-brain barrier, it will not gain access to the central nervous system to treat the disorder.

16.3 The type of transporters used by most drugs to enter human cells is best described as

  A. Active transporters.
  B. G proteins.
  C. GLUTs.
  D. Insulin dependent.
  E. Ion channels.

Correct answer = A. Drugs use active transporters to gain entry into human cells. G proteins function in cell signaling, and GLUTs are glucose transporters. Insulin is not required for drugs to enter cells, and drugs tend not to use ion channels to enter human cells.

16.4 A 24-year-old female is prescribed a β-lactam antibiotic to treat a bacterial infection. This drug will most likely cross intestinal epithelial cells to gain access to the circulation by

  A. ABC transporter–mediated transport.
  B. Binding to P-gp on the surface of epithelial cells.
  C. Inhibiting ATP hydrolysis of an ABC transporter.
  D. Passively diffusing through intestinal epithelial cells.
  E. PEPT transporter–mediated passage.

Correct answer = E. β-Lactam antibiotics are transported by PEPT transporters. ABC transporters transport drugs out of cells, not into them. Inhibition of ATP hydrolysis of an ATP transporter would not affect entry of a β-lactam antibiotic into intestinal epithelial cells. P-gp is an ABC transporter. Passive diffusion is no longer believed to be an important route of drug entry into cells.

16.5 A 38-year-old male has been taking St John's wort for 2 years. He obtains this herbal preparation over the Internet and has never mentioned his use of St John's wort to his physician. He develops atrial fibrillation (abnormal heart rhythm) and is prescribed digoxin. Therapy with digoxin is ineffective in treating his abnormal heart rhythm. This finding is best explained by

A. Competition of St John's wort and digoxin for the same SLC transporter.
B. MDR developing from St John's wort and digoxin.
C. Poor solubility of digoxin in the plasma membrane of intestinal epithelial cells.
D. St John's wort having greater affinity for a transporter than digoxin.
E. St John's wort up-regulation of P-gp that interferes with digoxin absorption.

Correct answer = E. St John's wort up-regulates P-gp, an ABC transporter. Overexpression of P-gp prevents the absorption of some other drugs, including digoxin, because P-gp exports the drug out of the cell. St John's wort and digoxin do not compete for binding to the same receptor, and receptor affinity would not be an important factor. Digoxin is normally transported across intestinal epithelial cells (when P-gp is not overexpressed), and its solubility in the membrane is not a consideration since a transporter is normally used.

16.6 A patient's cancer cells develop MDR to chemotherapy agents used in her treatment. Expression of which of the following proteins is most likely found on the cancer cells?

A. OATP
B. OATP1B1
C. ABCB1
D. PEPT
E. SLC

Correct answer = C. ABCB1 is the ABC transporter involved in MDR. It exports drugs from cells rendering the cells resistant to the effects of the drugs. SLCs are solute-linked carriers that transport many drugs into cells. OAT, OATB1P, and PEPT are examples of SLCs.

UNIT IV
Cell Signaling

Good communication is as stimulating as black coffee
and just as hard to sleep after.
—Anne Morrow Lindbergh (American author and aviator, 1906–2001)
In: *Gift from the Sea* (1955)

Soluble chemical signals, including hormones, growth factors, and neurotransmitters, sent from one cell to another, are a basic means by which cells communicate with each other. The cell that receives the signal is the target cell, and it binds the signaling molecule via a protein receptor on its surface or within its cytoplasm or nucleus. The binding of the signal to the receptor initiates a process that results in a cascade of reactions to amplify the signal and produce the desired effect within the cell. The types of receptors engaged by signaling molecules are grouped by their distinct cell signaling mechanisms, although a great deal of overlap exists in these classifications.

Biochemical processes within target cells are regulated in response to signaling molecules. The first cell signaling chapter focuses on G-protein signaling. Receptors coupled to G proteins amplify the message sent by the signaling molecule by regulating the production of intracellular signaling molecules including second messengers. The second chapter in this unit concerns signaling by catalytic receptors that possess enzymatic activity in the form of tyrosine kinases. The third chapter in the unit examines steroid hormone signaling, which is readily distinguished from the other two forms of signaling by the location of steroid hormone receptors in the interior of the cell as opposed to on the membrane's surface.

G protein and catalytic receptor signaling and steroid hormone signaling to some extent all involve phosphorylation of amino acid residues within cellular proteins by protein kinases. When serine or threonine amino acid residues are phosphorylated, cellular programs are turned on for hours or more. But the stimulatory effects of tyrosine phosphorylation are more fleeting, and a rapid cellular response ensues. Overstimulation of critical signaling pathways can cause heightened activation within a cell, with malignant consequences, and no rest for the cell if the inappropriate stimulation cannot be halted.

17 G-Protein Signaling

G proteins—Bind to GTP; Hydrolyze GTP to GDP	
Heterotrimeric G proteins	**Ras superfamily G proteins**
Three subunits, α, β, γ	Monomers resemble α subunit of heterotrimeric G proteins
Use G-protein linked receptors	Use catalytic receptors
Regulate second messengers	

Figure 17.1
Heterotrimeric and Ras superfamily G proteins.

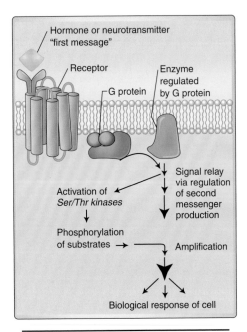

Figure 17.2
Overview of G-protein signaling.

I. OVERVIEW

G proteins are intracellular signaling proteins, named for their ability to bind to guanosine triphosphate (**GTP**). They also possess **GTPase** activity, which gives them the ability to hydrolyze GTP to GDP. Two categories of G proteins are described: **heterotrimeric G proteins** and the **Ras superfamily** of G proteins (Figure 17.1).

Ras superfamily members are often called "small G proteins" or "small GTPases" since they are monomers that resemble one subunit (α) of the heterotrimeric G proteins. Ras proteins receive their signals from catalytic receptors that have been activated by their ligand (see Chapter 18). The overall effects of Ras signaling often involve induction of cell proliferation, cell differentiation, or vesicle transport.

Heterotrimeric G proteins consist of three subunits, α, β, and γ. Signaling is initiated by ligand or hormone binding to receptors linked to G proteins tethered to the inner membrane leaflet. Activation of the G protein then enables it to regulate a specific membrane-bound enzyme. Products of reactions catalyzed by activated enzymes include **second messengers** that amplify the signal sent to the cell by the hormone or neurotransmitter that bound its receptor and acted as the first message (Figure 17.2). Many second messengers activate **serine/threonine protein kinases**, enzymes that phosphorylate their substrates on serine and threonine amino acid residues. Changes in phosphorylation status of target proteins, many of which are enzymes, can alter their activity. The overall result is the biological response of the cell to the hormone or neurotransmitter. The biological response is often the regulation of a biochemical pathway or the expression of a gene.

II. RECEPTORS AND HETEROTRIMERIC G-PROTEIN SIGNALING

Receptors for many hormones and neurotransmitters are linked to G proteins. G-protein–linked receptors are the most common form of cell surface receptor. These receptors have extracellular hormone-binding regions as well as intracellular portions that interact with the G protein to send the message from the hormone into the cell to evoke a response.

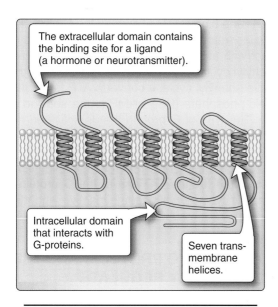

The extracellular domain contains the binding site for a ligand (a hormone or neurotransmitter).

Intracellular domain that interacts with G-proteins.

Seven trans-membrane helices.

Figure 17.3
Structure of G-protein–coupled receptors.

A. G-protein–linked receptors

G-protein–linked receptors are transmembrane proteins with seven membrane-spanning regions (Figure 17.3). Approximately 750 G-protein couple receptors are encoded by human genes, with about 350 of those known to bind to specific growth factors, hormones, or other known ligands. Most are expressed in multiple tissues. Over 90% of them are expressed in the brain.

B. Class of G-protein–linked receptors

This entire family of G protein–linked receptors has been traditionally categorized into three main classes, A, B, and C, based on distinct sequence homology among members of each group. Class A, the rhodopsin-like members, is the largest and includes olfactory receptors. A newer classification system involves six classes, A to F, based on both structure and function, and known as GRAFS (glutamate, rhodopsin, adhesion, frizzled, and secretin). Regardless of the category in which they are placed, all receptors linked to heterotrimeric G proteins use the same basic process to stimulate G proteins to regulate the production of second messengers.

C. Signaling mechanism

The basic mechanism for G-protein–linked signaling is shown for the G_s type of G protein (Figure 17.4). The process begins with an unoccupied G-protein–linked receptor that does not interact with the G protein that is in close proximity to its intracellular domain, such as the G_s shown here. Ligand binding to the receptor then creates an occupied receptor that undergoes a conformational change and is able to interact with the G protein. (A ligand is a molecule that binds specifically to a particular receptor. Hormones and neurotransmitters are ligands of G-protein–linked receptors.) In response to the receptor binding to the G-protein complex, the $G\alpha$ subunit of the G protein

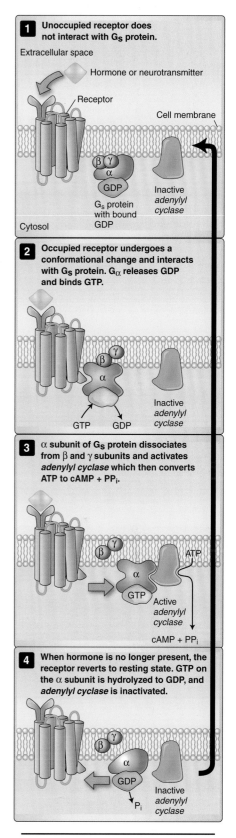

1 Unoccupied receptor does not interact with G_s protein.

Extracellular space

Hormone or neurotransmitter

Receptor

Cell membrane

β γ
α
GDP

Inactive *adenylyl cyclase*

G_s protein with bound GDP

Cytosol

2 Occupied receptor undergoes a conformational change and interacts with G_s protein. $G\alpha$ releases GDP and binds GTP.

β γ
α

Inactive *adenylyl cyclase*

GTP   GDP

3 α subunit of G_s protein dissociates from β and γ subunits and activates *adenylyl cyclase* which then converts ATP to cAMP + PP_i.

β γ
α
GTP

ATP

Active *adenylyl cyclase*

cAMP + PP_i

4 When hormone is no longer present, the receptor reverts to resting state. GTP on the α subunit is hydrolyzed to GDP, and *adenylyl cyclase* is inactivated.

β γ
α
GDP

P_i

Inactive *adenylyl cyclase*

Figure 17.4
Activation of G proteins.

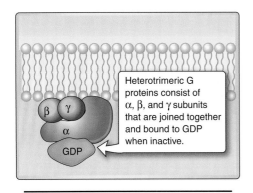

Figure 17.5
Heterotrimeric G proteins.

releases GDP and binds GTP, activating the G protein. The α subunit next dissociates from the β and γ subunits. The active α subunit goes on to interact with an enzyme whose function is regulated by the G protein. Adenylyl cyclase is the enzyme activated by the G_s type of G protein. Active adenylyl cyclase converts ATP to **cyclic AMP** (cAMP) and inorganic phosphate (PP_i). cAMP is the **second messenger** in G_s signaling. The type of G protein that is activated and the second messenger it regulates depend on the ligand, the type of receptor, and the type of target cell. When the ligand is no longer present bound to the receptor, the receptor will revert to its resting state. GTP will be hydrolyzed to GDP (by the GTPase of the G protein); the enzyme, such as adenylyl cyclase, will become inactive; and the α subunit will reassociate with β and γ subunits to stop the signaling process.

III. HETEROTRIMERIC G PROTEINS AND THE SECOND MESSENGERS THEY REGULATE

Distinct members of the heterotrimeric G-protein family exist through the association of various forms of the three subunits, α, β, and γ (Figure 17.5). At least 17 different α subunits are known in mammals and are grouped into four main categories. In all types of α subunit, GDP is bound to it when all three subunits are joined together in the inactive form. Certain Gα subunits interact with certain enzymes. For example, G_s interacts with adenylyl cyclase as described above. The four categories of Gα subunits include S, I, Q, and 12/13 are distinguished from each other by subscripts: $Gα_s$, $Gα_i$, $Gα_q$, and $Gα_{12}$, $Gα_{13}$. The identity of the enzyme determines which second messengers will be produced (or inhibited). Adenylyl cyclase and phospholipase C are two enzymes regulated by G proteins that are responsible for regulating messengers with important signaling roles.

A. Adenylyl cyclase

Two different Gα proteins regulate the activity of adenylyl cyclase; the $Gα_s$ system stimulates its activity while the $Gα_i$ inhibits it. Epinephrine (adrenaline) is a hormone that signals with cAMP as the second messenger. In liver, muscle, and adipose cells, the biological response that results is the breakdown of stored carbohydrates (glycogen) and fat for use as energy. Glucagon is a hormone that also stimulates glycogen breakdown in liver (see also *LIR Biochemistry*, pp. 131–134). In the heart, the number of beats per minute (heart rate) is increased by this signaling process.

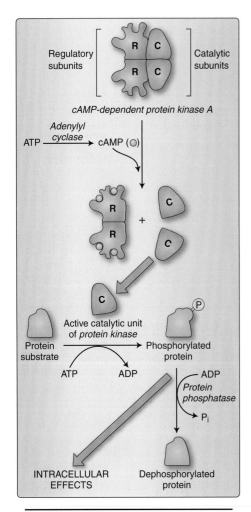

Figure 17.6
Activation of PKA by cAMP.

1. $Gα_s$: The active Gα stimulates adenylyl cyclase (see Figure 17.4). This enzyme uses ATP as a substrate to produce the second messenger cAMP. The enzyme **phosphodiesterase** converts cAMP to 5′-AMP, ensuring that the amount of cAMP in the cell is low. cAMP activates cAMP-dependent protein kinase A, known as **protein kinase A** (PKA) (Figure 17.6). The activation process involves cAMP binding to the regulatory or R subunits of PKA, enabling the release of catalytic or C subunits. Freed C subunits of PKA are active. PKA phosphorylates its protein substrates, many of which are enzymes, on serine and threonine residues. Phosphorylation regulates the activity of proteins and enzymes and can lead to intracellular effects. Protein phosphatases can dephosphorylate the phosphorylated proteins to regulate their activity. Over time, the $G\alpha_s$ will hydrolyze GTP to GDP to terminate the activation of adenylyl cyclase and the production of cAMP.

2. $G\alpha_i$: When $G\alpha_i$ is activated, it interacts with the active adenylyl cyclase to inhibit its ability to produce cAMP. In response, PKA will not be activated and its substrates will not be phosphorylated.

B. Phospholipase C

Phospholipase C is a family of enzymes that cleave membrane phospholipids. The family is divided by structure into six isoforms, β, χ, δ, ε, η, and ζ. Activation of each type of phospholipase C results in hydrolysis of the membrane phospholipid phosphatidylinositol 4,5-bisphosphate and formation of inositol 1,4,5-trisphosphate and diacylglycerol, which function as second messengers and described in more detail below.

1. $G\alpha_q$: A variety of neurotransmitters, hormones, and growth factors initiate phospholipase C activation through signaling via $G\alpha_q$ (Figure 17.7). After a hormone binds to its G_q-linked receptor, the intracellular domain of the occupied receptor interacts with G_q. The α subunit of G_q and $G_{q\alpha}$ releases GDP and binds GTP, and the α subunit then dissociates from the β and γ subunits. The now freed $G_{q\alpha}$ activates phospholipase C to cleave the membrane lipid phosphatidylinositol 4,5-bisphosphate (PIP_2). The products of this

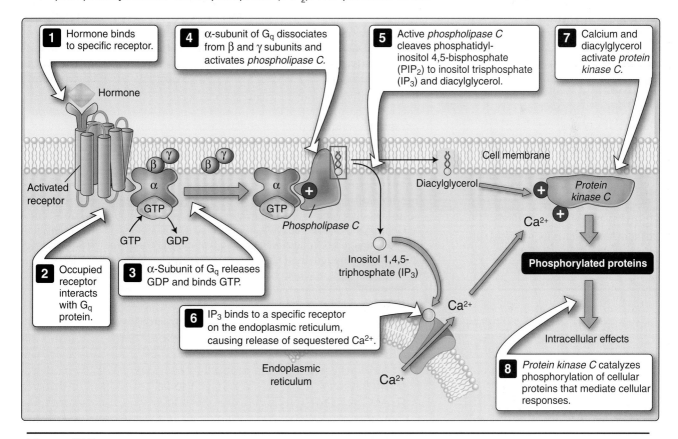

Figure 17.7
Generation of second messengers in response to $G\alpha_q$ activation of phospholipase C.

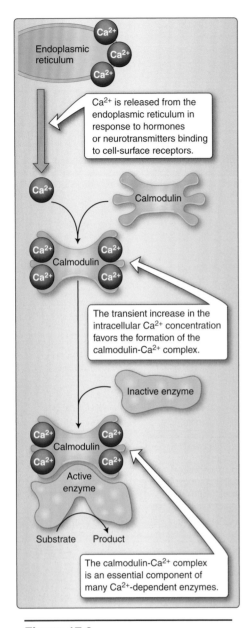

Figure 17.8
Calmodulin mediates many effects of intracellular calcium.

cleavage are **inositol 1,4,5-trisphosphate** (IP_3), which is released into the cytosol, and **diacylglycerol** (DAG), which remains within the plasma membrane. IP_3 binds to a specific receptor on the endoplasmic reticulum, causing release of sequestered calcium. Calcium and DAG together activate the calcium-dependent protein kinase named **protein kinase C** (PKC). **IP_3, DAG**, and **calcium** are **second messengers** in this system.

PKC catalyzes phosphorylation of cellular proteins that mediate cellular responses. Effects of intracellular calcium are mediated by the calcium-binding protein **calmodulin** (Figure 17.8). After calcium is released from the endoplasmic reticulum in response to the signaling of hormones or neurotransmitters, the transient increase in intracellular calcium concentration favors formation of the calmodulin-calcium complex. The calmodulin-calcium complex is an essential component of many calcium-dependent enzymes. Binding of the complex to inactive enzymes results in their conversion to active enzymes.

2. **$G\alpha_{12/13}$:** Members of the $G\alpha_{12/13}$ family are expressed in most cell types and can activate phospholipase C-ε. Similarities between $G_{q\alpha}$ and $G_{12/13\alpha}$ have been reported. Additionally, $G_{12/13\alpha}$ can activate phospholipase D and members of the Ras family of small GTPases. $G_{12/13}$ signaling is important in cell growth and in apoptosis. Disruptions in this pathway have been observed in leukemia cells and abnormal regulation may be involved in malignant cell transformation and metastasis.

Clinical Application 17.1: Toxins and Gα Proteins that Regulate Adenylyl Cyclase

Both cholera and pertussis toxins alter Gα subunits and higher than normal concentrations of cAMP in infected cells. Cholera toxin is produced by **Vibrio cholera** bacteria that produce cholera toxin when they infect intestinal epithelial cells. This toxin modifies the $G\alpha_s$b unit so that it cannot hydrolyze GTP and adenylyl cyclase remains active indefinitely. Diarrhea and dehydration result from excessive outflow of water into the gut in response to excess cAMP. Cholera can be fatal without appropriate hydration therapy. **Bordetella pertussis** is a bacterium that infects the respiratory tract and causes pertussis or whooping cough. Vaccination now prevents many young children from dying from the effects of pertussis. However, it remains a major health threat. The World Health Organization reports that there were 39 million cases and 297,000 deaths attributed to pertussis in 2000. Ninety percent of all cases are reported in developing countries, but the numbers of cases have been rising in the United States each year. This devastating disease is caused by pertussis toxin produced by the infecting bacteria. It inhibits $G\alpha_i$ so that $G\alpha_i$ cannot inhibit adenylyl cyclase. Adenylyl cyclase remains active indefinitely, producing excess cAMP. Coughing can lead to vomiting and dehydration. Antibiotics and hydration therapy are used in treatment.

IV. Ras G PROTEINS

Ras G proteins are homologous to the α subunits of heterotrimeric G proteins. They do not regulate membrane-bound enzymes or induce the production of second messengers. Instead, their activation by GTP allows them to initiate a cytoplasmic phosphorylation cascade that terminates with activation of gene transcription. In this signaling scheme, Ras proteins are viewed as relay switches between cell surface receptors and a cascade of serine/threonine kinases that regulate nuclear transcription factors. Such signaling is important in the regulation of cell proliferation. The aberrant function of Ras proteins may contribute to the malignant growth properties of cancer cells.

A. Signaling mechanism

Ras proteins are involved in signaling by certain hormones and growth factors that are ligands of catalytic receptors (see also Chapter 18). A linear pathway from the cell surface to the nucleus has been described, with Ras acting as an intermediary (Figure 17.9). Ligand

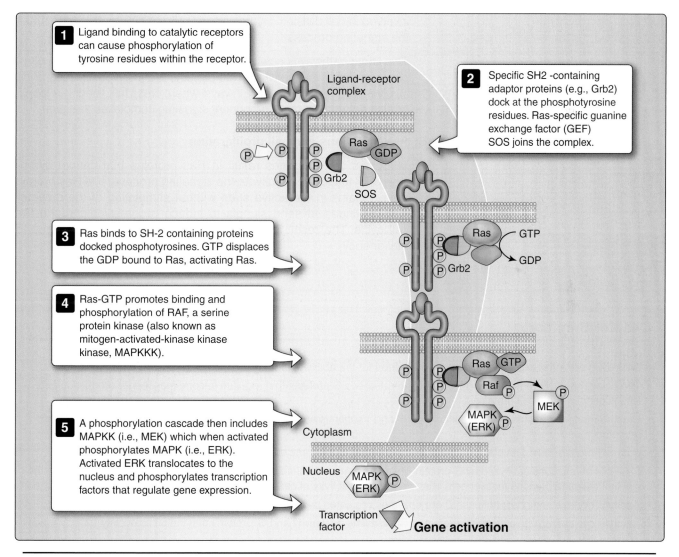

Figure 17.9
Ras signaling via activation of a cytoplasmic serine/threonine cascade.

binding to catalytic receptors can cause phosphorylation of tyrosine residues within the receptors. The receptor's phosphotyrosines provide "docking" or binding sites for intracellular adaptor proteins such as SHC and Grb2 that contain regions known as SH2 domains.

Ras-specific guanine exchange factor (GEF) SOS joins the complex, followed by Ras. The SHC-SOS-Ras complex exchanges GTP for GDP on Ras, activating Ras. Ras-GTP promotes binding and phosphorylation of Raf, a serine protein kinase (also known as MAPKKK for **m**itogen-**a**ctivated **p**rotein **k**inase **k**inase **k**inase). A phosphorylation cascade then includes mitogen-activated protein kinases kinases (such as MEK) that phosphorylate and activate mitogen-activated protein kinase (MAPK, also known as extracellular signal-regulated kinases or ERK), enabling it to translocate to the nucleus where it phosphorylates a transcription factor (such as ELK). The cascade terminates with transcription of genes for immediate early genes involved in cell division. Hydrolysis of GTP to GDP by Ras terminates the signaling process.

This linear pathway is now recognized to be only a part of a very complex signaling circuit in which Ras proteins are involved. Ras signaling involves a complex array of pathways, where cross talk, feedback loops, branch points, and multicomponent signaling complexes are seen.

B. Ras mutations and cell proliferation

Mutations in *Ras* genes result in Ras proteins that cannot hydrolyze GTP to GDP to inactivate the signaling process. The Ras protein then remains in the active state without stimulation of the receptor and continues to send signals to induce progression through the cell cycle. The result is excessive cell proliferation that can lead to malignancy.

Chapter Summary

- G proteins are intracellular signaling proteins named for the ability to bind to and hydrolyze GTP.
- Two categories of G proteins are described: heterotrimeric G proteins that regulate second messenger production and Ras superfamily small G proteins.
- Heterotrimeric G proteins are composed of α, β, and γ subunits and are activated by ligand binding to G-protein–linked receptors.
- Active G-protein–linked receptors interact with membrane-bound enzymes and regulate their function.
- Products of reactions catalyzed by G-protein–linked enzymes are second messengers that amplify the signal sent to the cell by the ligand. Second messengers often regulate the activity of certain serine/threonine protein kinases.
- Adenylyl cyclase and phospholipase C are enzymes regulated by G proteins.
- Adenylyl cyclase is regulated by G_s proteins that stimulate its activity and G_i proteins that inhibit its activity.
  - cAMP is the second messenger whose production is regulated by adenylyl cyclase.
  - cAMP activates PKA.

Chapter Summary (Continued)

- Phospholipase C is activated by G_q and $G_{12/13}$ proteins that stimulate its activity and allow it to cleave the membrane lipid PIP_2.
  - IP_3 and DAG are products of this cleavage and are the second messengers.
  - IP_3 induces the release of calcium from the endoplasmic reticulum.
  - Calcium and DAG activate PKC.
  - Calcium binds to calmodulin, which regulates the activity of other proteins.
- The GTP-binding protein Ras is an intermediary in signaling via some catalytic receptors.
- Activated Ras can stimulate the MAP kinase cascade of serine/threonine phosphorylations that can result in stimulation of gene transcription.
- Ras signaling is involved in the stimulation of cell proliferation. Mutations in *Ras* can cause unregulated cell division and malignancy.

Study Questions

Choose the ONE best answer.

17.1 Adenylyl cyclase is activated by a G protein. Which of the following second messengers will be generated?
   A. ATP
   B. cAMP
   C. Calcium
   D. DAG
   E. IP_3

Correct answer = B. cAMP is the second messenger generated by activated adenylyl cyclase that uses ATP as a substrate to produce the cAMP. Calcium, DAG, and IP_3 are generated in response to phospholipase C activation. PIP_2 is cleaved by phospholipase C to generate DAG and IP_3.

17.2 Manic-depressive illness may result from the overproduction of IP_3 and DAG and the accompanying signaling processes in certain CNS cells. Lithium is often useful in treating this illness. Lithium most likely functions to inhibit
   A. Adenylyl cyclase activity.
   B. $G\alpha_s$ protein function.
   C. Phospholipase C activity.
   D. PKA activity.
   E. Tyrosine kinase activity.

Correct answer = C. Phospholipase C is the enzyme regulated by G_q that catalyzes the production of IP_3 and DAG. Lithium's inhibition of phospholipase C inhibits the production of IP_3 and DAG. Adenylyl cyclase catalyzes the production of the second messenger cAMP when stimulated by active Gsα. PKA is regulated by cAMP. Tyrosine kinase activity is not involved in the production of DAG and IP_3.

17.3 A 6-month-old male patient presents with slight fever, rhinitis, and sneezing as well as forceful coughs ending with the loud inspiration (whoop). *Bordetella pertussis* was cultured from the nasopharynx. The toxin from this microorganism prevents the normal function of the $G\alpha_i$ protein in cells of the respiratory tract. Which of the following disruptions in cell signaling will result in the respiratory tract response to this infection?
   A. Calcium being unable to bind to calmodulin
   B. Impaired IP_3-stimulated release of calcium from endoplasmic reticulum
   C. Increased phospholipase C activity and PIP_2 cleavage
   D. Increased stimulation of PKC activity
   E. Overproduction of cAMP from uninhibited adenylyl cyclase

Correct answer = E. Overproduction of cAMP from uninhibited adenylyl cyclase will occur when G_i is inhibited by pertussis toxin. G_i normally inhibits adenylyl cyclase. Calcium is released in response to activation of phospholipase C. PKC is also activated as a consequence of phospholipase C activation.

17.4 Protein kinase A

    A. Activation by Ras stimulates gene transcription.

    B. Induces the release of calcium from endoplasmic reticulum.

    C. Is activated via G_q stimulation of phospholipase C.

    D. Phosphorylates protein substrates on serine/threonine residues.

    E. Stimulates the cleavage of PIP_2.

17.5 A constitutively overactive mutant form of *Ras* is present in cells from a breast biopsy sample. Therefore, *Ras* in these cells

    A. Acts as a serine/threonine kinase to terminate cell proliferation.

    B. Binds adenylate cyclase to overstimulate cAMP production.

    C. Catalyzes the breakdown of PIP_2.

    D. Is found in the nucleus bound to transcription factors.

    E. Overstimulates the MAP kinase cascade causing abnormal growth.

Correct answer = E. St John's wort up-regulates P-gp, an ABC transporter. Overexpression of P-gp prevents the absorption of some other drugs, including digoxin, because P-gp exports the drug out of the cell. St John's wort and digoxin do not compete for binding to the same receptor and receptor affinity would not be an important factor. Digoxin is normally transported across intestinal epithelial cells (when P-gp is not overexpressed), and its solubility in the membrane is not a consideration since a transporter is normally used.

Correct answer = E. Constitutively activated Ras will overstimulate the MAP kinase cascade and cause abnormal cell growth. Ras is not a protein kinase and does not bind to adenylyl cyclase. Ras does not participate in the PIP_2 signaling system. Ras is a cytoplasmic factor and does not enter the nucleus.

Catalytic Receptor Signaling

18

I. OVERVIEW

Growth factors, cytokines (growth factors of the immune system), and some hormones are signaling molecules that use catalytic or enzymatic receptors to stimulate their target cells (see also *LIR Immunology*, Chapter 6). Most catalytic receptors are single-chain transmembrane proteins that associate with other single-chain transmembrane proteins upon ligand binding and signal via phosphorylation of tyrosine (Tyr) residues. The **Tyr kinase** activity responsible for the production of phosphotyrosines may belong to the receptor itself or to a Tyr kinase that associates with the receptor. As a consequence of ligand binding to the receptor, intracellular portions of the receptor protein become phosphorylated on Tyr residues, then in a regulated manner, protein Tyr **phosphatases** rapidly dephosphorylate phosphotyrosines. The short-lived appearance of phosphotyrosines is a potent signal to the cell. Phosphorylated Tyr residues within the receptor induce binding of other proteins that serve as adaptors in the relay of the signal deeper into the cell. The original signal from the ligand may be split along several intracellular pathways as the signal is sent further onward with the goal of evoking a biological response to the ligand.

II. RECEPTORS WITH INTRINSIC TYROSINE KINASE ACTIVITY

Many **growth factors** signal via receptors with intrinsic Tyr kinase activity. Examples include transforming growth factor, epidermal growth factor, and platelet-derived growth factor (PDGFR). **Insulin**, a hormone, also signals via intrinsic catalytic receptors. Receptors with intrinsic Tyr kinase activity contain latent catalytic or enzymatic domains that are activated upon ligand binding. While many distinct catalytic receptors exist, they all share some common structural characteristics.

A. Receptor structure

Most catalytic receptors are formed by associations of two or more single transmembrane protein chains. Each of the single transmembrane chains has three domains: a ligand-binding portion that contains the amino-(NH_2)-terminus of the protein, an α-helical domain that spans the lipid bilayer, and an effector region that extends into the cytoplasm that contains the catalytic domain with Tyr kinase activity (Figure 18.1).

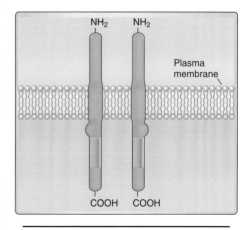

Figure 18.1
Structure of catalytic receptors.

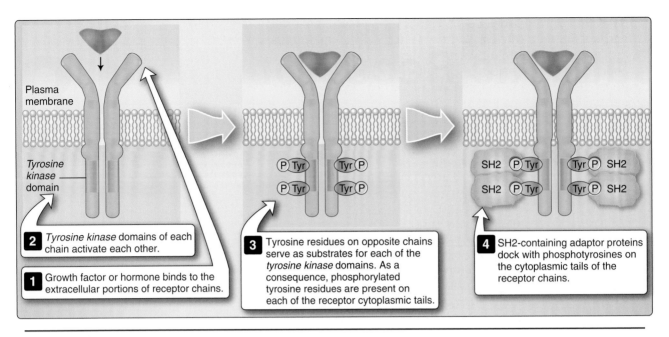

Figure 18.2
Initial steps in signaling by catalytic receptors with intrinsic tyrosine kinase activity.

B. Mechanism of signaling

In response to ligand binding, the individual transmembrane protein chains are brought physically closer together, often forming dimers (Figure 18.2). Tyr kinase domains of each receptor chain activate the other and the Tyr kinase within each receptor tail phosphorylates Tyr residues on intracellular portions of the other receptor chain. Tyr residues within the receptor itself serve as substrates for the receptor's Tyr kinase. This process is known as **autophosphorylation** since Tyr residues within the receptor protein are phosphorylated by enzymatic activity within the receptor itself. As a consequence, phosphotyrosine residues are present on each of the receptor cytoplasmic tails. Tyr phosphorylation triggers assembly of an elaborate intracellular signaling complex on the receptor tails (cytoplasmic domains). Intracellular proteins, known as **adaptor proteins**, which contain highly conserved domains (regions) known as SH2 and SH3 domains (named for their **S**rc **h**omology) dock with phosphotyrosines of the cytoplasmic tails of the receptor chains (Figure 18.3). Different receptors recruit distinct collections of SH2-containing adaptor proteins.

C. Important adaptor molecules

Adaptor molecules can function in signaling initiated by different growth factors or hormones. Certain adaptor molecules are known to be critically important in multiple signaling processes.

1. **Ras:** Ras is a small GTP-binding protein and serves as a molecular switch for key signaling in the control of growth and differentiation (see also Chapter 17, Figure 17.9). Ras does not contain an SH2 domain itself, but it binds to SH2-containing adaptor proteins that bind to phosphorylated Tyr residues within receptor tails. Ras activates a serine/threonine phosphorylation cascade called the

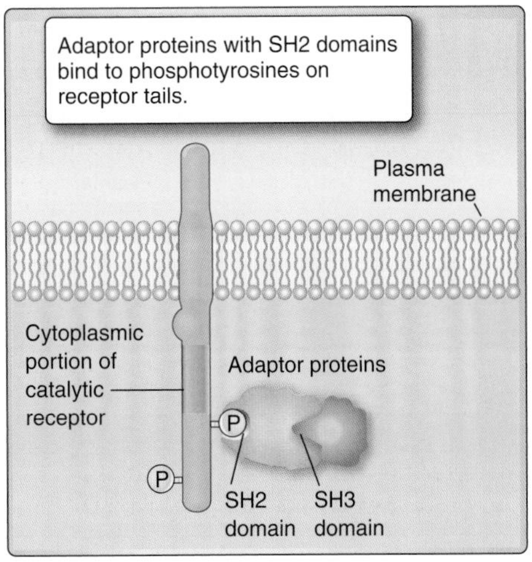

Figure 18.3
Binding of adaptor proteins.

MAP kinase cascade. Serine and threonine phosphorylations are longer lived than Tyr phosphorylations. The final enzyme in this cascade, mitogen-activated protein kinase (MAPK), becomes phosphorylated, translocates to the nucleus, and phosphorylates transcription factors. Phosphorylated transcription factors induce transcription of genes that allow the cell to proliferate or differentiate, depending on the nature of the signaling molecule at the cell's surface.

2. **STATs:** STATs are named for their function as signal transducers and activators of transcription. They are SH2-containing latent, cytoplasmic proteins that can bind to phosphorylated Tyr residues within the cytoplasmic domains of receptors with their own catalytic activity and are also involved in signaling by nonreceptor Tyr kinases. After ligand binding has induced receptor dimerization and Tyr phosphorylation of receptor tails, STATs dock with a phosphotyrosine on a receptor tail (Figure 18.4). When the STAT is bound to a phosphorylated Tyr residue, the Tyr kinase sees the STAT as a substrate and phosphorylates Tyr residues within the STAT protein. Tyr-phosphorylated STATs form dimers and translocate to the nucleus to bind to DNA and induce transcription of certain responsive genes.

D. PI3 kinase pathway

Another major signaling pathway stimulated by catalytic receptors is the phosphatidylinositol 3-kinase or PI3 kinase pathway that is important in promoting cell survival and cell growth. Following ligand binding to a catalytic receptor, receptor dimerization, and phosphorylation of Tyr residues within the cytoplasmic tail of the receptor, PI3 kinase binds

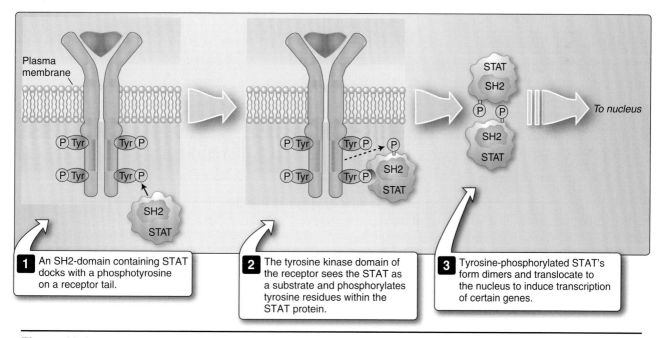

1 An SH2-domain containing STAT docks with a phosphotyrosine on a receptor tail.

2 The tyrosine kinase domain of the receptor sees the STAT as a substrate and phosphorylates tyrosine residues within the STAT protein.

3 Tyrosine-phosphorylated STAT's form dimers and translocate to the nucleus to induce transcription of certain genes.

Figure 18.4
STATs in signal transduction.

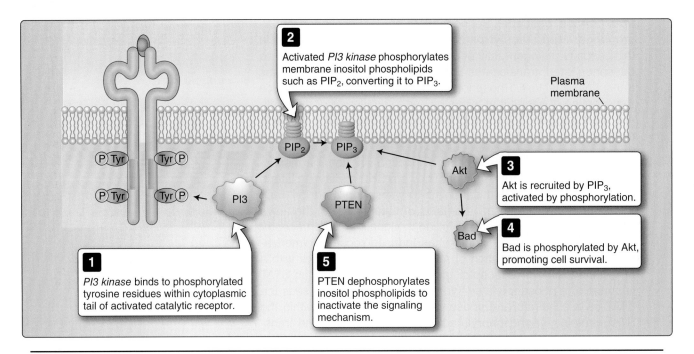

Figure 18.5
PI3 kinase pathway.

to the phosphorylated Tyr residues (Figure 18.5). Activated PI3 kinase phosphorylates membrane inositol phospholipids (on their three position), such as phosphatidylinositol 4,5-bisphosphate (PIP_2). PIP_2 is converted to PIP_3 in response to PI3 kinase action. Phosphorylated inositol lipids are docking sites for intracellular signaling proteins. **Akt**, also called protein kinase B, is recruited by PIP_3 and activated by phosphorylation. **Bad** is then phosphorylated by Akt, inactivating Bad and preventing it from inducing programmed cell death (apoptosis). Thus, cell survival is promoted. PIP_3 remains in the membrane until it is dephosphorylated by inositol phospholipid phosphatases, including **PTEN**, whose actions halt the signaling mechanism. (If a mutation occurs in PTEN, signaling through PI3 kinase can continue for prolonged periods and can promote cancer development.)

III. SIGNALING VIA NONRECEPTOR TYROSINE KINASES

Receptors for cytokines (interleukins and interferons) and for some hormones (such as prolactin and growth hormone) do not possess their own Tyr kinase activity but activate nonreceptor Tyr kinases to carry out their signaling process (see also *LIR Immunology*, pp. 72–75). The cytoplasmic domains of these receptors noncovalently associate with cytoplasmic Tyr kinase proteins and phosphorylate Tyr residues within the receptor tail. Several nonreceptor Tyr kinases have been identified, but two of the best characterized include the Src and Janus kinase families.

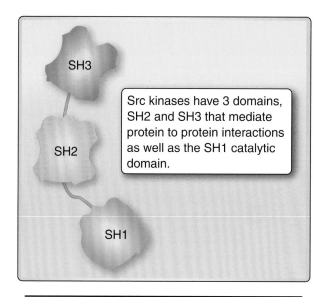

Figure 18.6
Src family of tyrosine kinases.

A. Src family tyrosine kinases

Src was the first nonreceptor Tyr kinase discovered. There are at least eight members of this nonreceptor protein Tyr kinase family including Blk, Fgr, Fyn, Hck, Lck, Lyn, Src, and Yes. All of these contain SH2 and SH3 domains that mediate protein to protein interactions as well as an SH1 catalytic domain (Figure 18.6). Different members of the family are found in different cell types. For example, Fyn, Lck, and Lyn function in lymphocyte signal transduction. Various members of the Src family can phosphorylate Tyr residues of many of the same target proteins. Src family members are regulated by phosphorylation on Tyr residues and by protein to protein interactions. Src proteins are normally inactive and are switched on only at crucial times. If Src remains active, uncontrolled growth and malignancy can result. Mutations in Src are found in many cancers.

B. Janus kinases

Janus kinases, referred to as **JAK**s, are latent cytosolic Tyr kinases activated by certain cytokine and hormone receptors. JAKs phosphorylate Tyr residues of the intracellular portion of the receptor chains. STATs bind to these phosphotyrosines and are phosphorylated on Tyr residues by the JAK (Figure 18.7). This signaling process is often referred to as the JAK-STAT pathway. Tyr-phosphorylated STATs form dimers and translocate to the nucleus as previously described.

IV. INSULIN SIGNALING

Insulin signals via catalytic receptors with intrinsic Tyr kinase activity. The insulin receptor is preformed in the membrane with all chains joined together prior to hormone binding. After insulin binds to the extracellular ligand-binding domain of its receptor, the insulin receptor Tyr kinase

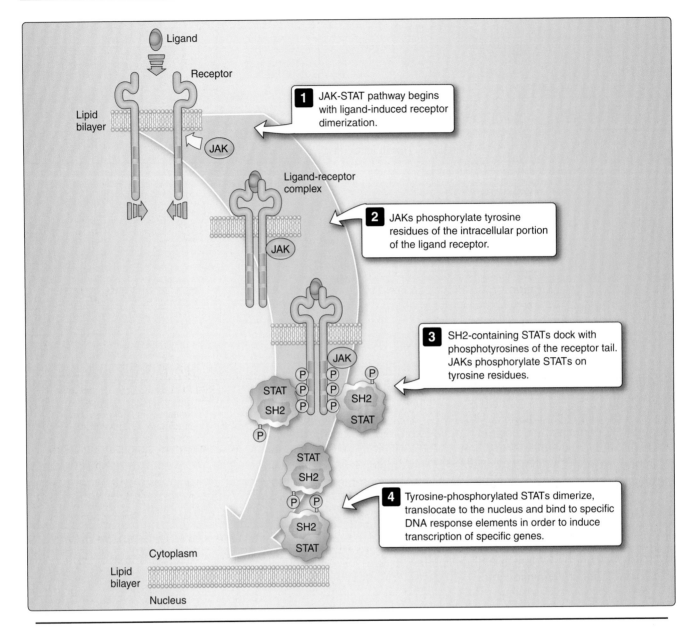

Figure 18.7
Janus kinase signaling.

activity is stimulated, inducing Tyr phosphorylation of various insulin receptor substrates (IRS) (Figure 18.8). At least four different IRS proteins are known. IRS-1 and IRS-2 are widely expressed; IRS-3 is found in adipose tissue, pancreatic β-cells, and possibly the liver; while IRS-4 is found in the thymus, brain, and kidney. Depending on the tissue type and the IRS proteins expressed, insulin will induce different biological responses to its signaling.

Recent research has shown that Tyr-phosphorylated IRS proteins activate several different intracellular signaling proteins including Ras, STATs, and PI3 kinase. This **signal splitting** occurs as the message sent

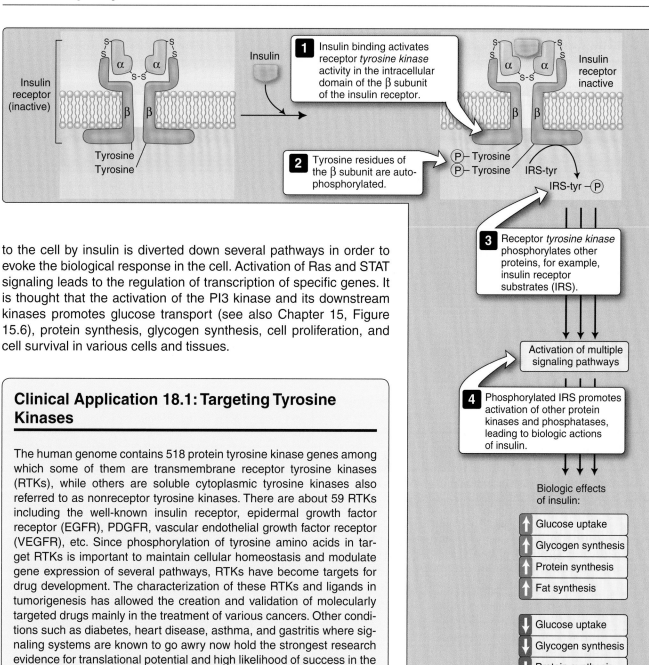

to the cell by insulin is diverted down several pathways in order to evoke the biological response in the cell. Activation of Ras and STAT signaling leads to the regulation of transcription of specific genes. It is thought that the activation of the PI3 kinase and its downstream kinases promotes glucose transport (see also Chapter 15, Figure 15.6), protein synthesis, glycogen synthesis, cell proliferation, and cell survival in various cells and tissues.

Clinical Application 18.1: Targeting Tyrosine Kinases

The human genome contains 518 protein tyrosine kinase genes among which some of them are transmembrane receptor tyrosine kinases (RTKs), while others are soluble cytoplasmic tyrosine kinases also referred to as nonreceptor tyrosine kinases. There are about 59 RTKs including the well-known insulin receptor, epidermal growth factor receptor (EGFR), PDGFR, vascular endothelial growth factor receptor (VEGFR), etc. Since phosphorylation of tyrosine amino acids in target RTKs is important to maintain cellular homeostasis and modulate gene expression of several pathways, RTKs have become targets for drug development. The characterization of these RTKs and ligands in tumorigenesis has allowed the creation and validation of molecularly targeted drugs mainly in the treatment of various cancers. Other conditions such as diabetes, heart disease, asthma, and gastritis where signaling systems are known to go awry now hold the strongest research evidence for translational potential and high likelihood of success in the clinic. Currently, the most exciting applications of tyrosine kinases have now progressed to the creation and application of second-generation drugs for cancers.

Figure 18.8
Insulin receptor autophosphorylation and IRS function.

Chapter Summary

- Most catalytic receptors are single-chain transmembrane proteins that form dimers when the ligand binds.
- Stimulation of phosphorylation of substrates on Tyr residues is a key feature of signaling by catalytic receptors.
- Some receptors have intrinsic Tyr kinase activity, while others associate with nonreceptor Tyr kinases.
- Receptors for growth factors and for the hormone insulin contain intrinsic Tyr kinase activity that is stimulated by ligand binding.
- Adapter molecules containing SH2 domains bind to the phosphorylated Tyr residues within the receptor cytoplasmic tails when the catalytic receptor is activated by its ligand. STATs are such adapter proteins that are activated to stimulate gene transcription.
- The PI3 kinase pathway promotes cell growth and survival and is stimulated by many catalytic receptors. Phosphorylation of inositol phospholipids stimulates additional signaling reactions.
- Src and Janus kinases are intracellular nonreceptor Tyr kinases.
- Insulin signals its catalytic receptor to undergo autophosphorylation on Tyr residues. The activated Tyr kinase domain of the receptor then phosphorylates various IRSs to send the signal onward in the cell.

Study Questions

Choose the ONE best answer.

18.1 Prolactin's stimulation of its target cell begins by its catalytic receptor

    A. Activating G proteins.
    B. Catalyzing production of second messengers.
    C. Dephosphorylating serine/threonine residues.
    D. Forming dimers in the membrane.
    E. Stimulating Tyr phosphorylation of Ras.

Correct answer = D. Dimerization of receptor chains within the membrane is an initial step of catalytic receptor signaling following ligand binding. Catalytic receptors do not activate G proteins or catalyze the production of second messengers. Dephosphorylation of serine/threonine residues is not a consequence of catalytic receptor signaling. Ras functions as a GTP-binding protein and does not become phosphorylated on Tyr residues.

18.2 Interleukin-2 binds to its catalytic receptor on a T lymphocyte. In response, which of the following will change within the cell?

    A. Activity of adenylyl cyclase
    B. Calcium concentration
    C. Level of phosphotyrosine
    D. Ras movement to the nucleus
    E. Second messengers

Correct answer = C. The level of phosphotyrosine within a cell will change in response to catalytic receptor signaling. Second messengers, including calcium, will not be altered by catalytic receptor signaling. Adenylyl cyclase is an enzyme linked to some G proteins. Its activity will not be affected by catalytic receptor signaling.

18.3 STATs function in signal transduction by

    A. Activating GTP binding to the α subunits of G proteins.
    B. Binding receptors phosphorylated on serine/threonine residues.
    C. Linking to G protein–coupled transmembrane receptors.
    D. Phosphorylating substrates on Tyr residues.
    E. Stimulating transcription of responsive genes.

Correct answer = E. STATs function in signal transduction by stimulating the transcription of responsive genes. STATs are first activated by phosphorylation of Tyr residues by either receptor Tyr kinases or JAK kinases. Tyr-phosphorylated STATs dimerize, translocate to the nucleus, and bind to DNA, stimulating transcription. STATs function independently of G proteins. They do not activate or bind to them. STATs do not bind to phosphorylated serine/threonine residues on receptors. STATs do not possess kinase activity and therefore do not phosphorylate substrates.

18.4 Insulin binds to an insulin receptor of an adipocyte (fat cell). Which of the following is a signaling process that will occur in response?

    A. Activation of protein kinase C to phosphorylate substrates

    B. Adenylyl cyclase stimulation of cAMP production

    C. G protein activation of second messenger production

    D. Translocation of the insulin receptor to the cell's nucleus

    E. Tyr phosphorylation of IRS

> Correct answer = E. Insulin signaling involves phosphorylation of IRS on Tyr residues. Insulin signaling does not activate the serine/threonine kinase, protein kinase C, which results from second messenger activation. Second messengers, such as cAMP, are not utilized in insulin signaling. The insulin receptor remains embedded in the plasma membrane and does not translocate to the cell's nucleus.

18.5 A cell has a mutant form of PTEN. As a result, which of the following signaling molecules will remain active longer than usual?

    A. G protein

    B. JAK kinase

    C. PI3 kinase

    D. Protein kinase C

    E. STATs

> Correct answer = C. PI3 kinase will remain active when mutant PTEN is unable to dephosphorylate inositol phospholipids. G protein signaling does not involve PTEN. Protein kinase C activity is stimulated by G protein–stimulated second messengers and does not depend upon PTEN. Both JAK kinases and STATs function independently of PTEN.

19 Steroid Receptor Signaling

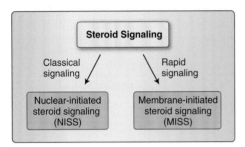

Figure 19.1
Mechanisms of steroid signaling.

Type I Receptors

Sex hormone receptors
(androgen, estrogen, progesterone receptors)

Glucocorticoid receptor

Mineralocorticoid receptor

Type II Receptors

Vitamin A receptor

Vitamin D receptor

Retinoid receptor

Thyroid hormone receptor

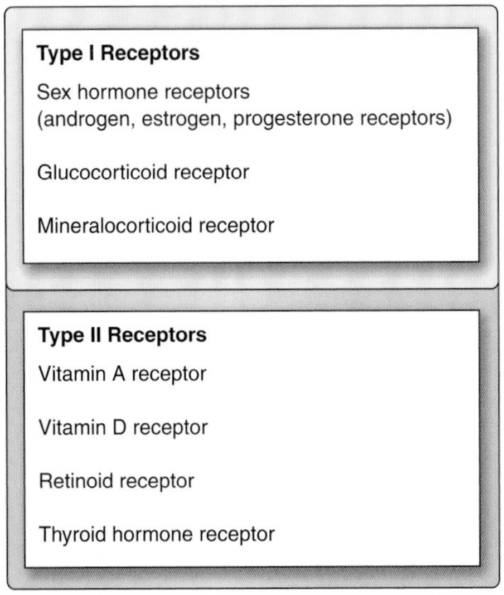

Figure 19.2
Categories of steroid receptors.

I. OVERVIEW

The use of intracellular receptors distinguishes classical steroid hormone signaling from signaling by hydrophilic signaling factors including peptide hormones and growth factors that use extracellular, membrane-bound receptors. Intracellular receptors for steroids are located in the cytoplasm or in the nucleus of target cells. Steroid hormone receptors act as **ligand-activated transcription factors**, since their ligand (hormone) binds to them and activates them so that they can bind to DNA and regulate transcription (production of mRNA) for a specific gene, which is then translated into a protein. This classical form of steroid signaling is known as **nuclear-initiated steroid signaling** (NISS) (Figure 19.1). In this way, the steroid has an effect on the cell's genome. It can take minutes, hours, or days for the effects of classical steroid hormone signaling to induce a biological response in the target cell in the form of production of a new protein.

Within seconds or minutes after addition of some steroid hormones, some other signaling effects can be observed in target cells. These include changes in intracellular calcium concentration, activation of G proteins, and stimulation of protein kinase activity, which are not mediated by classical intracellular steroid receptors but through steroid receptors in the plasma membrane. **Membrane-initiated steroid signaling** (MISS) is now described in addition to classical steroid signaling. This more rapid, membrane-initiated form of steroid signaling is also referred to as **nongenomic actions of steroid hormones**. At present, less detailed information is known about MISS than about NISS. Both types of signaling are believed to be important for normal function of steroid hormones.

Molecules that signal target cells using NISS and MISS include sex steroid hormones, glucocorticoids, and mineralocorticoids as well as vitamins A and D, retinoids, and thyroid hormones. Classical intracellular receptors have been grouped into two categories, type 1 receptors and type 2 receptors based on the details of their signaling mechanisms (Figure 19.2). Sex hormone, glucocorticoid, and mineralocorticoid receptors are type 1 receptors, while vitamin A, vitamin D, retinoid, and thyroid hormone receptors are type 2. In order to bind to and activate their intracellular receptors, the steroid hormones and vitamins must first move from the blood circulation across cell membranes. Steroid hormones are synthesized from a common precursor and have structures that enable them to enter into their target cells.

II. STEROID HORMONES

Cholesterol is the precursor of all classes of steroid hormones: gluco-corticoids (e.g., cortisol), mineralocorticoids (e.g., aldosterone), and sex hormones—androgens, estrogens, and progestins (Figure 19.3). (Note: Glucocorticoids and mineralocorticoids are collectively called corticosteroids.) Cholesterol is first converted to pregnenolone and then to progesterone, which is a common precursor to all steroid hormones. Corticosteroids such as cortisol and aldosterone are produced from pro-gesterone. While testosterone (an androgen) is also produced from pro-gesterone, estradiol (an estrogen) is produced from testosterone (see *LIR Biochemistry*, Chapter 18 for more information regarding cholesterol).

Synthesis and secretion of steroid hormones occur in the adrenal cortex (cortisol, aldosterone, and androgens), ovaries and placenta (estrogens and progestins), and testes (testosterone) (Figure 19.4). Hormones exert their effects at the cellular level, as evidenced by aldosterone stimulation of renal reabsorption of sodium and excretion of potassium. Other biolog-ical effects of steroid hormones include cortisol's stimulation of gluconeo-genesis, estrogen's regulation of the menstrual cycle, and testosterone's promotion of anabolism.

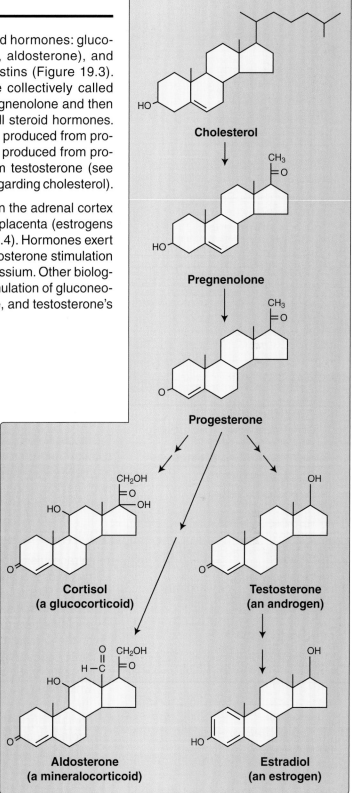

Figure 19.3
Key steroid hormones produced from cholesterol.

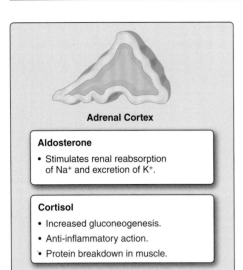

Adrenal Cortex

Aldosterone
- Stimulates renal reabsorption of Na⁺ and excretion of K⁺.

Cortisol
- Increased gluconeogenesis.
- Anti-inflammatory action.
- Protein breakdown in muscle.

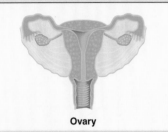

Ovary

Estrogens
- Control menstrual cycle.
- Promote development of female secondary sex characteristics.

Progesterone
- Secretory phase of uterus and mammary glands.
- Implantation of maturation of fertilized ovum.

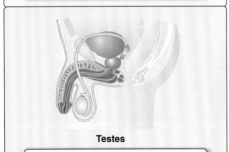

Testes

Testosterone
- Stimulates spermatogenesis.
- Promotes development of male secondary sex characteristics.
- Promotes anabolism.
- Masculinization of the fetus.

Figure 19.4
Actions of steroid hormones.

In order to exert such biological consequences, steroid hormones are transported by the blood from their sites of synthesis to their target organs. Because of their lipid nature and hydrophobicity, they must be complexed with a plasma protein in the aqueous environment of blood plasma. Plasma albumin can act as a nonspecific protein carrier and does carry aldosterone. However, specific steroid-carrier plasma proteins bind the steroid hormones more tightly than does albumin, for example, corticosteroid-binding globulin (transcortin) is responsible for transporting cortisol and sex hormone–binding protein transports sex steroids.

Clinical Application 19.1: Inhibitors of Steroid Hormone Synthesis as Cancer Therapy

Estrogen is derived from testosterone by the action of the enzyme aromatase. Aromatase inhibitors are used in the treatment of estrogen-responsive breast cancer in postmenopausal women. After menopause, the main source of estrogen is from aromatization of adrenal-produced androgens. Inhibitors of aromatase can reduce estrogen levels significantly and remove the main source of growth stimulation from estrogen-responsive tumors. Arrest of tumor growth and/or initiation of apoptosis (programmed cell death) of estrogen-responsive breast tumors occur as a result of therapy with aromatase inhibitors.

III. NUCLEAR-INITIATED STEROID SIGNALING

In classical steroid signaling, NISS, steroid hormones must leave the circulation and cross the plasma membrane of a target cell. Once inside the cell, they will encounter a specific receptor in the cytosol or in the nucleus. Hormone binding modifies the receptor, enabling it to regulate the transcription of specific genes.

A. Intracellular receptor structure

Intracellular receptors for steroid hormones are a highly conserved group of proteins that contain three major functional domains (Figure 19.5). The **hormone (or ligand)-binding domain** is in the COOH-terminal region of the receptor protein while the NH₂-terminal region contains the **gene regulatory domain**. The **DNA-binding domain** of the protein forms an additional functional region. This region is highly conserved and contains **zinc finger motifs** containing cysteine amino acid residues that bind zinc and dictate DNA sequences to which the receptor will bind. Since these receptor proteins must enter the nucleus in order to bind to DNA and regulate transcription, they contain nuclear localization signals (NLS) to permit their trafficking into the nucleus (see also Chapter 11, Figure 11.10 for more information on trafficking of proteins into the nucleus).

B. Mechanism of nuclear-initiated steroid signaling

In the absence of hormone, estrogen and progesterone receptors are principally located in the nucleus of the target cell, and glucocorticoid and androgen receptors are located in the cytoplasm. Receptors for vitamins A and D, retinoids, and thyroid hormone (type 2 steroid receptors) are found in the nucleus (see also Figure 19.2). Regardless of the receptor's intracellular location, binding of a steroid hormone to its intracellular receptor causes activation of the receptor and enables it to translocate to the nucleus (Figure 19.6). The steroid hormone-receptor complex binds to the hormone response element (HRE) of the enhancer region and activates the gene promoter, causing transcription.

1. **Sex steroid receptors, glucocorticoid receptors, and mineralocorticoid receptors:** The activated receptor-ligand complex associates with coregulator or coactivator proteins that promote transcription. The receptor-ligand-coregulator complex binds to regulatory DNA sequences called HREs through zinc finger motifs. Ligand-bound type 1 receptor complexes bind to DNA as homodimers (two identical ligand-receptor complexes binding together). Binding of the activated hormone-receptor complexes to an HRE positions the activated receptor so that its gene regulatory domain interacts with proteins of the transcriptional complex bound to a promoter.

2. **Vitamins A and D, retinoid, and thyroid hormone receptors:** For these type 2 steroid receptors, unoccupied receptors are complexed in the nucleus with corepressor proteins, inhibiting them from inducing transcription. Ligand binding to the receptor causes release of the corepressor proteins and allows for binding to coactivator proteins. Other type 2 receptors form heterodimers with the retinoid X receptor when binding to DNA to regulate the transcription of vitamin- or hormone-responsive genes.

C. Hormone specificity of gene transcription

An HRE is found in the promoter (or an enhancer element) for genes that respond to a specific steroid hormone, thus ensuring coordinated regulation of these genes. For example, a glucocorticoid response element or GRE allows for a transcriptional response to a glucocorticoid such as cortisol. Each of the cortisol-responsive genes is under the control of its own GRE. Binding of the receptor-hormone complex to the glucocorticoid receptor (GR) causes a conformational change in the receptor that uncovers its zinc finger DNA-binding domain (Figure 19.7). The steroid-receptor complex then interacts with specific regulatory DNA sequences and the hormone-receptor complex in association with coactivator proteins controls the transcription of targeted genes. Overall, this process allows for the coordinate expression of a group of target genes, even when these genes are located on different chromosomes. The GRE can be located upstream or downstream of the genes it regulates and is able to function at great distances from those genes. The GRE, then, can function as a true enhancer.

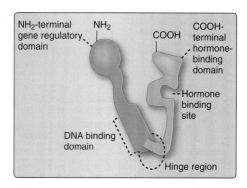

Figure 19.5
Structure of steroid hormone receptors.

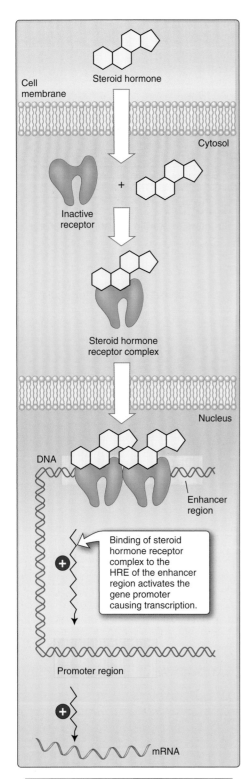

Figure 19.6
The nuclear-initiated steroid signaling (NISS) mechanism involves the activation of transcription by interaction of steroid hormone-receptor complex with hormone response element (HRE).

The label within the figure reads: Binding of steroid hormone receptor complex to the HRE of the enhancer region activates the gene promoter causing transcription.

Clinical Application 19.2: Hormone Receptor Antagonists as Cancer Therapies

Receptor antagonists bind to the hormone receptor and prevent binding of the natural hormone to its receptor. Selective estrogen receptor modulators (SERMs) are important therapies for treatment and prevention of breast cancer. Owing to their selectivity, SERMs have different effects in different tissues. One, tamoxifen, blocks estrogen receptors in breast, thereby inhibiting estrogen-dependent growth of tumors. Tamoxifen is used in premenopausal women with estrogen receptor–positive breast cancer. Tamoxifen has other effects in other tissues. For example, it can increase estrogen signaling in the endometrium, with the potential for endometrial malignancy.

IV. MEMBRANE-INITIATED STEROID SIGNALING

Rapid effects of steroid hormones that occur within seconds to minutes after exposure of target cells to steroid hormones are now believed to result from actions of steroid receptors localized to the plasma membrane. MISS induces biological effects more quickly than classical NISS since it promotes modifications to existing proteins (e.g., phosphorylation) and does not require synthesis of new proteins. Details remain to be determined for many aspects of MISS; however, some details of this signaling process are known, particularly for membrane estrogen receptors.

Membrane forms of androgen, glucocorticoid, progesterone, mineralocorticoid, and thyroid hormones have also been identified and have similar signaling processes to the membrane estrogen receptor. Evidence also exists for the presence of membrane-bound vitamin D receptors. Cross talk between intracellular and membrane-bound pools of steroid receptors is believed to occur, and both NISS and MISS mechanisms can be used to evoke biological responses. Convergence of signals at the membrane, cytoplasm, and nucleus causes the overall biological effects of steroid hormones. For example, kinases activated by MISS may phosphorylate coactivators required for transcriptional activation via NISS. Additionally, signaling from membrane steroid receptors may contribute to gene transcription independently of nuclear steroid receptors.

A. Membrane receptors

Membrane steroid receptors are believed to have the same protein structure as intracellular steroid receptors but are localized to membrane caveolae, invaginated regions of the membrane that have a flask-like shape (Figure 19.8) (see also Chapter 3, Figure 3.13). In its membrane-bound form, the steroid hormone receptor may either associate with the outer surface of the plasma membrane in the flask of an individual caveola or may be tethered by a scaffolding protein to the plasma membrane. After the receptor is bound by its specific steroid hormone, it becomes activated and may form homodimers or heterodimers with other membrane steroid receptors.

B. Mechanism of membrane-initiated steroid signaling

Activated receptors bound by their specific steroid hormone then associate with a complex of signaling proteins that can include G proteins,

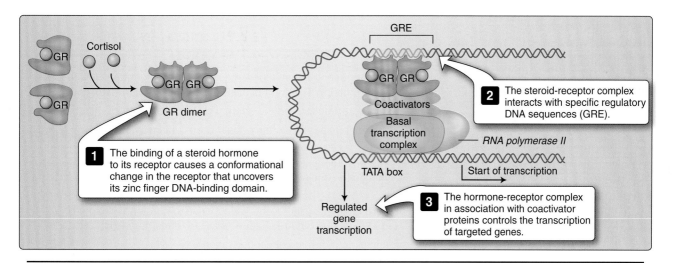

Figure 19.7
Transcriptional regulation by intracellular steroid hormone receptors. GRE, glucocorticoid response element (an example of an HRE); GR, glucocorticoid receptor.

growth factor receptors, the tyrosine kinase Src, and the GTP-binding protein Ras. The epidermal growth factor receptor (EGFR) is often implicated in MISS, and its activation can result in sustained MAP kinase signaling in a cell responding to a steroid via MISS. Second messengers can be induced and ion channels regulated. Protein kinases that often function in response to G protein activation, including serine/threonine kinases PKA and PKC, can be activated as well as PI3 kinase that functions in catalytic receptor signaling (see also Chapters 17 and 18). Phosphorylation of target proteins by activated kinases causes a rapid change in their activity and a rapid biological response by the cell.

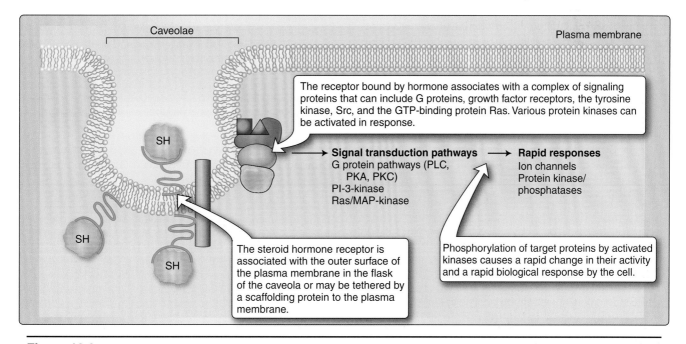

Figure 19.8
Membrane-initiated steroid signaling (MISS). SH, steroid hormone.

Chapter Summary

- Classical steroid signaling involves the use of intracellular receptors that function as ligand-activated transcription factors and therefore regulate the synthesis of new cellular proteins. This form of steroid signaling is referred to as NISS.

- Steroid hormones are synthesized from cholesterol and exert actions on the adrenal cortex, the ovaries, and the testes.

- Intracellular steroid receptors are present in the cytosol or nucleus of the target cell. They contain ligand-binding, gene regulatory, and DNA-binding domains.

- Hormone binding to its intracellular receptor results in receptor dimerization and activation. If in the cytosol, hormone-receptor complexes will first traffic to the nucleus. Once in the nucleus, they bind to DNA and activate transcription of hormone-responsive genes.

- Membrane receptors for steroid hormones permit more rapid signaling events in response to hormone binding. These have the same protein structure as intracellular steroid receptors but are instead localized to membrane caveolae.

- MISS involves modifications to existing cellular proteins, often through phosphorylation.

- Convergence of steroid signaling pathways at the membrane, cytoplasm, and nucleus permits an overall biological response to the steroid hormone.

Study Questions

Choose the ONE best answer.

19.1 A particular cell type under study is known to be the target of a certain hormone's action. The hormone has an intracellular receptor that normally resides within the cell's nucleus. The most likely identity of the hormone is
  A. Adrenocorticotropic hormone.
  B. Estrogen.
  C. Growth hormone.
  D. Insulin.
  E. Prolactin.

Correct answer = B. Estrogen is a steroid hormone that has an intracellular receptor that normally resides in the cell's nucleus. Adrenocorticotropic hormone (ACTH), growth hormone, insulin, and prolactin are all hydrophilic, peptide hormones that exclusively use membrane-bound receptors to stimulate their target cells.

19.2 Cells isolated from a breast tumor and grown in cell culture in the laboratory are found to respond to estrogen via both intracellular and membrane-bound receptors. Which of the following is most likely to result from signaling via the intracellular receptor?
  A. Activation of serine/threonine protein kinase
  B. Change in intracellular calcium levels
  C. GTP binding to Ras
  D. Phosphorylation of cellular enzymes
  E. Synthesis of estrogen-responsive protein

Correct answer = E. In response to an intracellular steroid receptor, a target cell will synthesize a hormone-responsive protein. In this situation, estrogen will regulate the transcription of an estrogen-responsive gene that will be translated into a protein. Changes in intracellular calcium levels and in serine/threonine kinase activity result from G protein–mediated cell signaling. G protein–linked receptors are membrane-bound receptors. Phosphorylation of substrates is catalyzed by protein kinases. Serine/threonine kinases are regulated by second messengers and G protein–linked receptors. Tyrosine kinases are activated in catalytic receptor signaling. All use membrane-bound receptors. Ras is a GTP-binding protein activated by some forms of membrane receptor signaling.

19.3 A 43-year-old female patient has an estrogen receptor–positive breast tumor. An SERM is used as treatment. The beneficial actions of this drug to the patient result from the drug
  A. Activating the transcription of estrogen-responsive genes.
  B. Augmenting estrogen signaling in the remaining tumor cells.
  C. Blocking the binding of estrogen to estrogen receptors.
  D. Inducing signaling via membrane estrogen receptors.
  E. Stimulating the translation of estrogen-regulated proteins.

Correct answer = C. Selective estrogen receptor modulators are hormone receptor antagonists that block the binding of the natural hormone to its receptor. Blocking the binding prevents hormone signaling through the receptor. In the case of an estrogen-responsive breast tumor, blocking the binding of estrogen will prevent the cancer cells from receiving stimulation necessary for their continued growth and survival. Activation of transcription of estrogen-responsive genes, augmentation of estrogen signaling, and stimulation of translation of estrogen-regulated proteins would be beneficial to the cancer cells and their survival, but not to the patient. It would not be beneficial to the patient for membrane-bound estrogen receptors to be stimulated and that is not the goal of such a therapy.

19.4 Which of the following regions within an androgen receptor protein contains zinc finger motifs?

   A. Cytosolic domain
   B. DNA-binding domain
   C. Gene regulatory domain
   D. Ligand-binding domain
   E. Transmembrane domain

Correct answer = B. The DNA-binding domain of an intracellular steroid receptor protein contains zinc finger motifs that bind zinc and dictate DNA sequences to which the receptor will bind. The ligand-binding domain binds to the hormone and is in the COOH-terminal region. The gene regulatory domain is important for transcriptional activation but does not contain the zinc fingers that actually facilitate DNA binding. Membrane-bound steroid receptors, including those for androgens, may contain cytosolic (extracellular) and transmembrane regions; however, neither of them would have zinc fingers to facilitate DNA binding.

19.5 A steroid-responsive tumor cell has abnormally increased activity of MAP kinase signaling in response to the steroid. Which of the following aspects of steroid signaling may be involved in this abnormally sustained signaling?

   A. DNA binding by activated hormone-receptor complex
   B. Formation of hormone-receptor complex
   C. Membrane-initiated steroid signaling
   D. Stimulation of the receptor's gene regulatory domain
   E. Translocation of the receptor to the cell's nucleus

Correct answer = C. Abnormal MISS can result in sustained activation of cellular kinase systems, including MAP kinase. EGFR is implicated in this abnormal signaling pathway. All other answer choices relate to NISS. Kinases such as MAP kinase are not (directly) stimulated in the course of NISS.

UNIT V

Regulation of Cell Growth and Cell Death

Life is pleasant. Death is peaceful. It's the transition that's troublesome.

—Isaac Asimov (American science fiction writer and biochemist, 1920–1992)

Arguably, the most important occurrences within the life of a cell are its generation from a progenitor and then, at the end of its life span, its demise, through a natural or a pathological process. Regulation of both cell generation and cell death is critical to ensure that the appropriate number of cells is available to function within the organism. Safeguards are also needed to protect against uncontrolled growth, which may result in malignancy and the death of the whole organism.

This unit begins with a description of the cell cycle, the orderly sequence of biochemical events that culminates with the generation of two new cells from one parent cell. Cells that enter the active phases of the cell cycle often do so after resting in interphase for varying periods of time, depending on the cell type. Once a cell commits itself to dividing and to passing on its genetic information, it enters into a critical transition period, which, if unsuccessful, will mean that the cell has failed to reproduce itself and will likely die. Curiously, if the outcome of the cell cycle is a success, the parent cell will cease to exist, with two exact replicas replacing it.

The second chapter in this unit concerns regulation of the cell cycle, and the third focuses on abnormal cell growth. Checks and balances within the cell cycle are necessary to enable the orderly process of cell duplication to occur. As our understanding of abnormal cell growth progresses, improved treatments to halt unregulated growth may be developed. The fourth chapter in this unit focuses on cell death, particularly on the physiological process of apoptosis, where collateral damage is minimized. This unit concludes with the final chapter, an exploration of aging and senescence of the cell and the organism.

20 The Cell Cycle

I. OVERVIEW

Multicellular organisms are composed of a variety of specialized cells that are organized into a cellular community. When an organism requires additional cells, either for growth or to replace damaged or aged cells, new cells must be produced by **cell division**, or **proliferation**. Somatic cells are generated by the division of existing cells in an orderly sequence of events. They duplicate their contents and then divide to produce two identical **daughter cells**. This sequence of duplication is known as the **cell cycle** and is the essential mechanism of eukaryotic reproduction.

Cell division occurs throughout the life of the organism, although different cell types divide more or less often than others. Cells display remarkable variation in their proliferative capacity, depending on the cell type and the age of the individual. For example, fibroblasts derived from neonates can complete close to 50 rounds of division, but fibroblasts isolated from adults can complete only approximately half as many cell cycles.

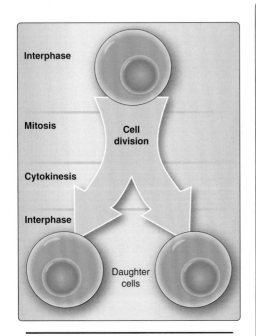

Figure 20.1
Stages of cell division and the cell cycle.

Clinical Application 20.1: Cell Renewal

Homeostasis, or stable system maintenance, requires that as cells die or are lost (e.g., through abrasion or sloughing), they are replaced by cells specific to that tissue. Cellular turnover is a normal function. The turnover times for some cells in the adult body are slow or nonexistent, in the endocrine and central nervous system, for example, whereas other cells turnover very rapidly. Each adult human has approximately 2×10^{13} erythrocytes. Because the half-life for an erythrocyte is approximately 115 days, the human body must replace approximately 10^{11} new red blood cells every day! The most abundant leukocytes, neutrophils, have a half-life of approximately 10.5 hours, which means the body needs to replace approximately 6×10^{10} neutrophils per day. Cells within the epithelia also turnover rapidly. The life span for cells lining the stomach is 3 to 5 days and for enterocytes that line the small bowel is 5 to 6 days.

For a cell to generate two daughter cells, complete copies of all of the cell's constituents must be made. Genetic information contained within multiple chromosomes must be duplicated; the cytoplasmic organelles and cytoskeletal filaments must be copied and shared between the two newly formed daughter cells.

The cell cycle may be broadly divided into three distinct stages: interphase, mitosis, and cytokinesis. **Interphase** is the period between successive rounds of nuclear division and is distinguished by cellular growth and new synthesis of DNA. It can be further subdivided into three phases called **G$_1$ phase**, **S phase**, and **G$_2$ phase** (Figure 20.1). Division of genetic information occurs during the stage known as **mitosis**, which can be divided into five distinct phases called **prophase**, **prometaphase**, **metaphase**, **anaphase**, and **telophase**. Mitosis assures that each daughter cell will have identical, complete, functional copies of the parent cell's genetic material. The third stage, cytoplasmic division or **cytokinesis**, culminates with the separation into two distinct daughter cells that enter interphase.

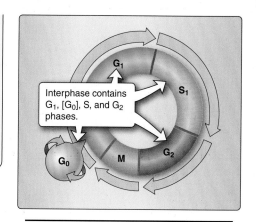

Figure 20.2
Interphase.

II. INTERPHASE

All cells, whether actively cycling or not, spend the vast majority of their lives in interphase. **Interphase** is an eventful and important part of the cell cycle and is composed of G$_1$, S, and G$_2$ phases (Figure 20.2). Cellular growth and DNA synthesis occur during interphase, resulting in a duplication of cellular contents so that there are sufficient materials for two complete new daughter cells.

A. G$_1$ and G$_0$ phases

Named for the gap that follows mitosis and the next round of DNA synthesis, **G$_1$ phase** is both a growth phase and a preparation time for DNA synthesis in S phase (Figure 20.3). RNA and protein synthesis also take place during G$_1$. In addition, organelles and intracellular structures are duplicated and the cell grows during this phase. The length of G$_1$ is the most variable among cell types. Very rapidly dividing cells, such as growing embryonic cells, spend very little time in G$_1$. On the other hand, mature cells that are no longer actively cycling are permanently in G$_1$. Those cells in G$_1$ that are not committed to DNA synthesis are in a specialized resting state called **G$_0$** (pronounced G-zero). Some inactive or quiescent cells in G$_0$ phase may re-enter the active phases of the cell cycle upon proper stimulation. The **restriction point** is located within G$_1$ phase and, if passed, will commit a cell to continuing into DNA synthesis within S phase. The restriction point is critical for cell cycle regulation and is detailed in Chapter 21.

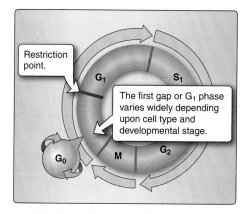

Figure 20.3
G$_1$ and G$_0$ phase.

B. S phase

Synthesis of nuclear DNA, also known as DNA **replication**, occurs during S phase (Figure 20.4). Each of the 46 chromosomes in a human cell is copied to form a sister **chromatid**. ATP-dependent unwinding of the chromatin structure by **DNA helicase** exposes the binding sites for DNA polymerase that will catalyze the synthesis of new DNA in the 5′ to 3′ direction. Multiple replication forks are activated on each chromosome in order to ensure that the entire genome is duplicated within the time span of S phase. Upon completion of DNA synthesis, chromosome strands are condensed into tightly coiled heterochromatin. The time for completion of this process is relatively constant among cell types. Actively cycling

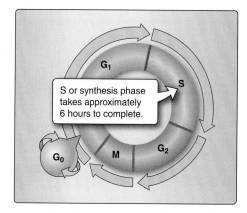

Figure 20.4
S phase.

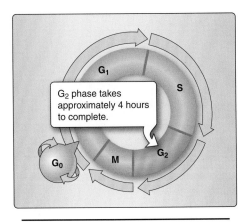

Figure 20.5
G$_2$ phase.

cells spend approximately 6 hours in S phase. DNA replication is described in detail in Chapter 7.

C. G$_2$ phase

The gap between the completion of S phase and the start of mitosis, known as **G$_2$**, is a time of preparation for the nuclear division of mitosis (Figure 20.5). This safety gap allows the cell to ensure that DNA synthesis is complete and correct before proceeding to nuclear division in mitosis. Also contained with G$_2$ is a checkpoint where intracellular regulatory molecules assess nuclear integrity (see Chapter 21.III). Typically, this phase lasts for approximately 4 hours.

III. MITOSIS

Mitosis, nuclear division, is a continuous process that can be divided into five descriptive phases based on progress made in the overall nuclear division. Dividing cells spend about 1 hour in mitosis (Figure 20.6). After completion of nuclear division in mitosis, cytokinesis, division of the cytoplasm, occurs. At the conclusion of cytokinesis, two separate daughter cells have been formed.

A. Prophase

In prophase, the nuclear envelope remains intact while the chromatin that was duplicated during S phase condenses into defined chromosomal structures called **chromatids** (Figure 20.7A). Chromosomes of mitotic cells contain two chromatids connected to each other at a **centromere**. Specialized protein complexes, called **kinetochores**, form and associate with each chromatid. Mitotic spindle microtubules will attach to each kinetochore as chromosomes are moved apart later in mitosis. The microtubules of the cytoplasm disassemble and then reorganize on the surface of the nucleus to form the **mitotic spindle**. Two centriole pairs push away from each other by growing bundles of microtubules forming the mitotic spindle. The **nucleolus**, the organelle within the nucleus where ribosomes are made, disassembles in prophase.

B. Prometaphase

The disassembly of the nuclear envelope marks the beginning of prometaphase (Figure 20.7B). Spindle microtubules bind to kinetochores and chromosomes are pulled by the microtubules of the spindle.

C. Metaphase

Metaphase is characterized by chromatids aligned at the equator of the spindle, halfway between the two poles (Figure 20.7C). The aligned chromatids form the metaphase plate. Cells can be arrested in metaphase when microtubule inhibitors are used (see also Chapter 21). Karyotype analyses performed to determine overall chromosome composition and structure most often require cells in metaphase.

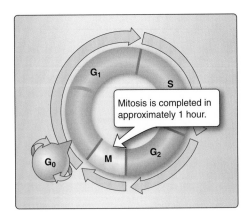

Figure 20.6
M phase.

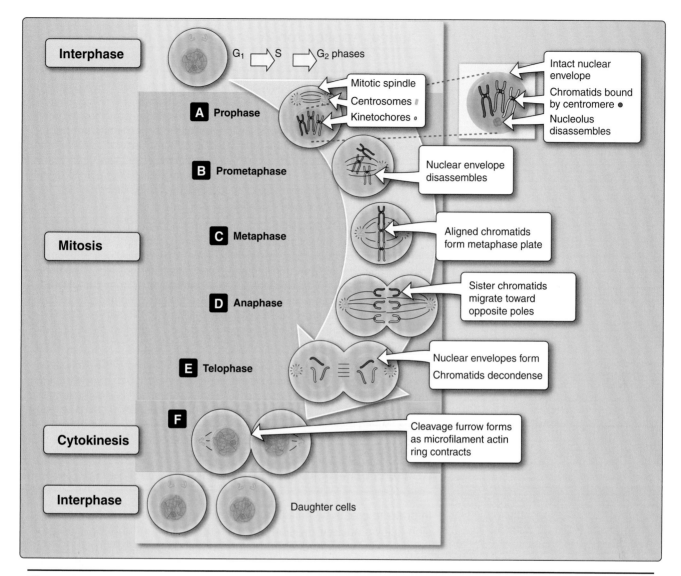

Figure 20.7
Mitosis. For illustrative purposes, three chromosomes have been drawn.

D. Anaphase

In anaphase, the mitotic poles are pushed further apart as a result of polar microtubules elongating (Figure 20.7D). Each centromere splits in two and paired kinetochores also separate. Sister chromatids migrate toward the opposite poles of the spindle.

E. Telophase

The last phase of nuclear division, telophase, is characterized by kinetochore microtubule disassembly and mitotic spindle dissociation (Figure 20.7E). Nuclear envelopes form around each of the two nuclei containing the chromatids. The chromatids decondense into dispersed chromatin and nucleoli reform in the daughter nuclei.

Clinical Application 20.2: Hayflick Limit

Early in the 20th century, researchers observed that cancerous tumors arising in rodents could be serially transplanted into other rodents indefinitely. By midcentury, cancer cells were shown to be immortal in tissue culture. In the 1960s, Dr. Leonard Hayflick made the startling observation that normal, noncancerous cells have a limited replicative capacity and are mortal. Hayflick discovered that human umbilical cord fibroblasts stopped dividing after about 50 divisions in culture—a phenomenon that has come to be known as Hayflick limit. Fibroblasts cultured from adults were able to divide far fewer times. Replicative capacity depends upon the number of cell divisions and not on the age of the cells.

Contributing to Hayflick limit is the irreversible shortening of each chromosome's telomere (a hexameric DNA repeat sequence, TTAGGG, at the end of each human chromosome) each time a cell divides. Although telomerase, a complex of RNA and protein, helps maintain and repair telomeres by adding telomeric repeats, material is eventually lost from the telomere, contributing to cellular senescence or aging. Telomeres normally help move chromosomes to opposite poles within the cell during telophase. When telomeres become too short, chromosomes can no longer segregate and the cell can no longer divide.

Some tissues require continuous cell replacement, such as skin, gut epithelia, and erythrocytes. These cells derive from progenitor stem cells that do not exhibit Hayflick limit. Other cells that are subject to Hayflick limit rarely divide, such as cells of the endocrine system, or never divide, such as neurons, during adult life.

IV. CYTOKINESIS: COMPLETION OF A CELL CYCLE

In order to create two distinct, separate daughter cells, cytoplasmic division follows nuclear division. An actin microfilament ring forms to create the machinery needed. Contraction of this actin-based structure results in the formation of a **cleavage furrow** that is seen beginning in anaphase (Figure 20.7F). The furrow deepens until opposing edges meet. Plasma membranes fuse on each side of the deep cleavage furrow, and the result is the formation of two separate daughter cells, each identical to the other and to the original parent cell.

Clinical Application 20.3: Aurora Kinases

First discovered in the eggs of the African clawed toad *Xenopus laevis*, aurora kinases are a family of serine/threonine kinases that play important functional roles during mitosis, specifically by controlling chromatid segregation. Three members of the aurora kinase family have been discovered in mammalian cells. Aurora A functions in prophase and is critical for proper formation of the mitotic spindle and for recruitment of proteins to stabilize centrosomal microtubules. Without Aurora A, the centrosome does not accumulate sufficient γ-tubulin for anaphase and the centrosome never fully matures. Aurora A is also necessary for proper separation of centrosomes after the spindle has been formed. Aurora B functions in the attachment of the mitotic spindle to the centromere and also in cytokinesis for cleavage furrow formation. Aurora C is a component of a key regulatory complex in mitosis called the chromosomal passenger complex. This complex ensures that chromosomes are aligned and segregated correctly and is needed for microtubule spindle assembly. Elevated expression of all three members of the aurora kinase family has been observed in many human tumors. Inhibitors of aurora kinases have been assessed as anticancer therapies. However, limited efficacy has been seen in clinical trials against solid tumors. One explanation for the lack of growth inhibition with these inhibitors is that the rate of cell proliferation in solid tumors is often quite slow. Hematopoietic malignancies appear to be more susceptible to growth inhibition by these potential therapeutic agents, since their rate of growth tends to be much higher than in solid tumors. The use of aurora kinase inhibitors in conjunction with other anticancer agents may be beneficial.

V. ASSESSMENT OF THE CELL CYCLE

Assessments of cell proliferation and of the cell cycle are both clinically important for the evaluation of tumor progression. Equally important to both cell biology and drug discovery research are methods used to evaluate cell proliferation and the role of agents that promote or slow the cell cycle. Although there are a number of tools and methods to assess proliferation, they can basically be divided into those used to analyze cell proliferation and those used to assess the cell cycle.

A. Assessment of cell proliferation

The proliferation of cells may be assessed either by measuring the synthesis of new (nascent) DNA or by the serial dilution of labeled cytoplasmic proteins as the cell divides.

1. **DNA synthesis:** Replication of DNA may be assessed using modified analogs of thymidine, one of the nucleoside building blocks of DNA. In one experimental approach, labeled or tagged thymidine or an analog of thymidine (e.g., BrdU) is added into tissue culture medium in which cells are grown. Because thymidine is exclusively used for DNA synthesis, cells that are actively synthesizing DNA will incorporate the labeled thymidine or analog, which can be measured (Figure 20.8).

2. **Dilution of a cytoplasmic probe:** Cytoplasmic probes can also be used to assess cell proliferation. With this approach, cells are incubated with the succinimidyl ester of carboxyfluorescein diacetate (CFSE), which readily crosses plasma membranes and enters the cytoplasm. There, intracellular esterases cleave the acetate groups making the compound both fluorescent and membrane impermeant, thus trapping CFSE within the cell. The succinimidyl ester groups of CFSE readily and irreversibly bind to available amines (usually on lysine) on intracellular cytoplasmic and membrane proteins. As cells divide, their fluorescently labeled cytoplasmic proteins are divided equally between the two daughter cells. Each daughter cell has half the fluorescence of the previous generation, which can be measured by flow cytometry (Figure 20.9).

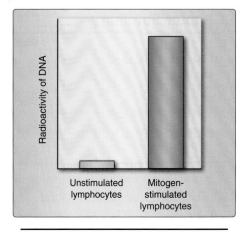

Figure 20.8
Cell proliferation assessed by ³H-thymidine incorporation by stimulated lymphocytes.

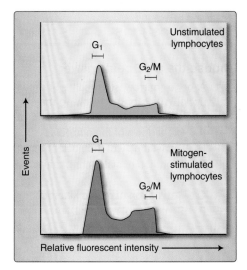

Figure 20.9
Cell proliferation by stimulated lymphocytes assessed by CFSE.

B. Cell cycle analysis

The amount of DNA contained within a cell is dependent on the phase of the cell cycle and ranges between $1n$ in G_1 and $2n$ in G_2 and M phases. Distribution of cells in a population within different cell cycle phases can be assessed by flow cytometry, to evaluate treatment therapies in lymphomas and leukemias and as a research tool in evaluating oncogene and tumor suppressor gene mechanisms. Any one of a wide variety of nucleic acid–binding fluorescent dyes may be used to label DNA. Fluorescence is proportional to the DNA content of the cell. Analysis of a flow cytometry histogram shows the proportion of cells within the population in G_1, S, and G_2 phases of the cell cycle (Figure 20.9).

Chapter Summary

- New cells to account for growth or to replace cells lost by injury or disease are produced from existing somatic cells by cell division.
- The sequence of duplication and division is known as the cell cycle.
- The cell cycle is divided into three stages: interphase, mitosis, and cytokinesis.
- All cells, including those cycling actively, spend the majority of their time in interphase, which is composed of G_1, S, and G_2 phases.
- Interphase is an eventful phase that includes cellular growth, protein and RNA synthesis in G_1 phase, DNA synthesis in S phase, and preparation for mitosis during G_2 phase.
- In mitosis, nuclear division follows interphase and culminates in two separate nuclei identical to each other and to the nucleus of the parent cell that was copied.
- Phases within mitosis include prophase, prometaphase, metaphase, anaphase, and telophase.
- Cytokinesis, or cytoplasmic division, occurs after mitosis and results in two distinct, separate daughter cells identical to each other and to the parent cell.

Study Questions

Choose the ONE best answer.

20.1 A bone marrow stem cell is in interphase of the cell cycle. Which of the following may be observed in this cell?

A. Breakdown of the nucleolus

B. Disassembly of the nuclear envelope

C. Migration of sister chromatids toward opposite poles

D. Separation of paired kinetochores

E. Synthesis of nuclear DNA

Correct answer = E. Synthesis of nuclear DNA occurs during S phase, one of the three phases that constitutes interphase. G_1 and G_2 are the other phases of interphase. Breakdown of the nucleolus occurs in prophase of mitosis. Disassembly of the nuclear envelope occurs in prometaphase of mitosis. Both migration of sister chromatids toward opposite poles and separation of paired kinetochores occur in anaphase of mitosis.

20.2 A hepatocyte that is actively participating in the cell cycle is observed to increase in size and to duplicate its organelles during a particular phase. In which phase does this hepatocyte presently reside?
   A. G_1 phase
   B. G_2 phase
   C. Prophase
   D. S phase
   E. Telophase

Correct answer = A. G1 phase is characterized by increases in cellular size and duplication of organelles, prior to replication of nuclear DNA in S phase. G2 phase is a safety gap prior to nuclear division of mitosis. Prophase and telophase are both phases of mitosis.

20.3 A mitotic cell is described as residing in telophase. Which of the following may be observed in this cell?
   A. Chromosome alignment at the equator
   B. Cleavage furrow formation
   C. Mitotic spindle dissociation
   D. RNA and protein synthesis
   E. Unwinding of the chromatin

Correct answer = C. Mitotic spindle dissociation and kinetochore microtubule disassembly characterize telophase. Chromosome alignment at the equator occurs in metaphase of mitosis. Cleavage furrow formation occurs during cytokinesis, which happens after the completion of mitosis. RNA synthesis and protein synthesis are seen in cells in G_1 phase of interphase, while unwinding of chromatin occurs in S phase of interphase.

20.4 An actively cycling lymphocyte is observed to have a separation of its paired kinetochores and a migration of its sister chromatids toward opposite poles of the spindle. This mitotic lymphocyte presently resides in
   A. Anaphase.
   B. Metaphase.
   C. Prometaphase.
   D. Prophase.
   E. Telophase.

Correct answer = A. Anaphase is characterized by splitting of two of the centromeres and separation of paired kinetochores along with the migration of sister chromatids to opposite poles of the spindle.

20.5 A cell that is stimulated to divide lacks Aurora A. This cell will be unable to complete
   A. Anaphase.
   B. G_1 phase.
   C. Metaphase.
   D. Prometaphase.
   E. S phase.

Correct answer = A. Without Aurora A, the centrosome does not accumulate sufficient γ-tubulin for anaphase and the centrosome never fully matures. Phases of mitosis that occur prior to anaphase (prophase, prometaphase, and metaphase) as well as phases within interphase (G_1, S, and G_2) can occur normally without Aurora A.

21

Regulation of the Cell Cycle

I. OVERVIEW

Numerous checks and balances ensure that the cell cycle is highly regulated, establishing a state of balance or **homeostasis** between cell proliferation, cell differentiation, and cell death. Certain cell types retain the ability to divide throughout their life spans. Others permanently leave the active phases of the cell cycle ($G_1 \rightarrow S \rightarrow G_2$) after their differentiation. Yet, other cells exit and then reenter the cell cycle. According to their type and function, cells receive developmental and environmental cues and respond in accord.

Cells that temporarily or reversibly stop dividing are viewed as being in a state of **quiescence** in the G_0 **phase** (Chapter 20) (Figure 21.1).

Distinct from temporarily resting quiescent cells, **senescent** cells have permanently stopped dividing, either due to age or due to accumulated DNA damage. For example, neurons are considered senescent and will not reenter the active phases of the cell cycle. (See Chapter 24 for further discussion of senescence.) Intestinal epithelial cells and bone marrow hematopoietic cells, on the other hand, undergo continuous, rapid cell turnover in the course of their normal function and must be continuously replaced. Liver hepatocytes do not continuously traverse the cell cycle but retain the ability to do so if needed. This ability of hepatocytes to reenter the active cell cycle accounts for liver regrowth following injury or disease; a property that has been successfully exploited in live-donor liver transplantation where portions of the liver from a donor are given to a patient in need of a liver transplant. Within several weeks after surgery, the liver tissue doubles in size in both the donor and the recipient.

II. CELL CYCLE REGULATORS

Cell cycle regulators control progression through the various stages of the cell cycle. Cell cycle mediators are categorized as **cyclins** or as **cyclin-dependent kinases (CDKs)**. The patterns of expression of these proteins and enzymes depend upon the cell cycle phase. Complexes of certain cyclins with specific CDKs **(cyclin-CDKs)** possess enzymatic (kinase) activity. Whenever necessary, **cyclin-dependent kinase inhibitors** (CKI) can be recruited to inhibit cyclin-CDK complexes (Figure 21.2).

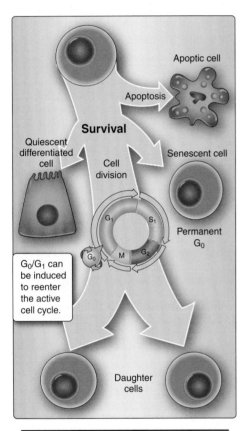

Figure 21.1
Tissue homeostasis requires a balance between differentiation, cell growth, and cell death.

Table 21.1: Cell Cycle Function of Cyclins and CDKs

Cyclin	Kinase	Function
D	CDK4 CDK6	Progression past the restriction point at the G$_1$/S boundary
E, A	CDK2	Initiation of DNA synthesis in early S phase
B	CDK1	Transition from G2 to M

A. Cyclins

The cyclins are a family of cell cycle regulatory proteins that are categorized as D, E, A, or B cyclins, which are expressed to regulate specific phases of the cell cycle. Cyclin concentrations rise and fall throughout the cell cycle due to synthesis and degradation (via the proteosomal pathway, Chapter 12) (Figure 21.3).

D-type cyclins (cyclins D1, D2, and D3) are G$_1$ regulators that are critical for progression through the **restriction point**, the point beyond which a cell will irrevocably proceed through the remainder of the cell cycle. S phase cyclins include type E cyclins and cyclin A (Table 21.1). Mitotic cyclins include cyclins B and A.

B. Cyclin-dependent kinases

CDKs are serine/threonine kinases that are present in constant amounts during the cell cycle. However, their enzyme activities fluctuate depending upon available concentrations of cyclins required for CDK activation (Figure 21.3). The specific cyclin binds first to the CDK and then **CDK-activating kinase** (CAK) phosphorylates the CDK on a threonine residue, completing its activation. Next, the active **cyclin-CDK complex** catalyzes the phosphorylation of substrate proteins on serine and threonine amino acid residues. Phosphorylation changes the activation status of substrate proteins. Such alteration of regulatory proteins allows for initiation of the next phase of the cell cycle.

Active CDK2 is responsible for activating target proteins involved in movement from G$_1$ to S (S phase transition) and for initiation of DNA synthesis. CDK1 targets activated proteins critical for the initiation of mitosis.

III. CHECKPOINT REGULATION

Checkpoints placed at critical points in the cell cycle monitor the completion of critical events and, if necessary, delay the progression to the next stage of the cell cycle (Figure 21.4). One such checkpoint is the **restriction point** in G$_1$. Before it reaches the restriction point, a cell requires external, growth factor stimulation to progress through G$_1$. After that, the cell will continue through the cell cycle without the need for further stimulation. Another is the **G$_2$ checkpoint**, as described below. The **S phase checkpoint** includes monitoring of cell cycle progression and

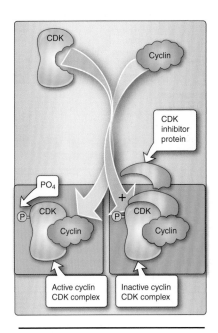

Figure 21.2
Cell cycle mediators and their formation of complexes.

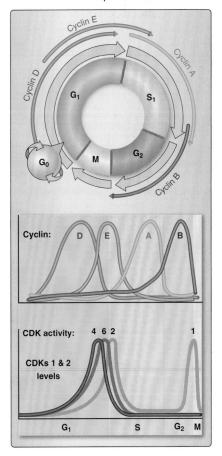

Figure 21.3
Cell cycle–specific expression of cyclins and activation of CDKs.

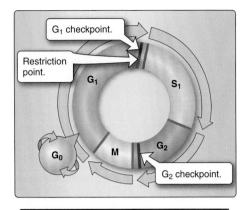

Figure 21.4
Progression through the cell cycle requires cyclin activation of specific CDKs.

if DNA damage occurs to an S-phase cell, the rate of DNA synthesis will be slowed to attempt to allow for repair.

A. G_1 checkpoint

It is important that nuclear synthesis of DNA not begin until all the appropriate cellular growth has occurred during G_1. Therefore, there are key regulators that ensure that G_1 is completed prior to the start of S phase, including tumor suppressors and CDK inhibitors. Tumor suppressor proteins normally function to halt cell cycle progression within G_1 when continued growth is not needed or is undesirable or when DNA is damaged. Mutated versions of tumor suppressor genes encode proteins that permit cell cycle progression at inappropriate times. Cancer cells often show mutations of tumor suppressor genes.

1. **Retinoblastoma (RB) protein:** The tumor suppressor **RB** normally halts cells in the G_1 phase of the cell cycle. When RB is mutated, such as in an inherited eye malignancy known as hereditary retinoblastoma, the cell is not stopped in G_1 and continues unregulated progression through the remainder of the cell cycle.

   In normal, **resting cells**, the RB protein contains few phosphorylated amino acid residues. In this state, RB prevents a cell's entry into S phase by binding to transcription factor E2F and its binding partner DP1/2 which are critical for the G_1/S transition (Figure 21.5). Therefore, RB normally prevents progression out of early G_1 and into S phase in a resting cell.

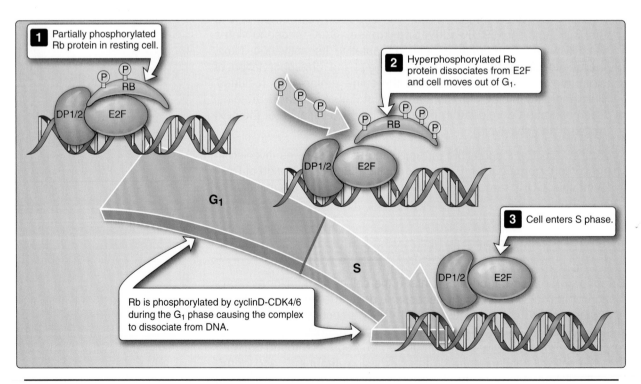

Figure 21.5
RB activation of transcription of S-phase genes.

In **actively cycling cells**, RB is progressively hyperphosphorylated as a consequence of growth factor stimulation and signaling via the MAP kinase cascade (Chapter 17). Subsequently, cyclin D-CDK4/6 complexes are activated and they phosphorylate RB. Further phosphorylation of RB by cyclin E-CDK2 allows the cell to move out of G_1. Hyperphosphorylated RB can no longer inhibit transcription factor E2F binding to DNA. Therefore, E2F is able to bind to DNA and activate genes whose products are important for S phase. Examples of E2F-regulated genes include thymidine kinase and DNA polymerase, both of which are involved in the synthesis of DNA.

2. **p53:** The **p53** tumor suppressor protein plays a major regulatory role in G_1. When DNA is damaged, p53 becomes phosphorylated, stabilized, and activated. Activated p53 stimulates transcription of **CKI** (see Figure 21.2) to produce a protein named **p21**, to halt cell cycle progression to allow for DNA repair. If the DNA damage is irreparable, p53 instead triggers apoptosis (Chapter 23).

If p53 is mutated and unable to arrest the cell cycle, then unregulated cell cycle progression can occur. Over 50% of all human cancers show p53 mutations (Figure 21.6).

3. **Cyclin-dependent kinase inhibitors:** Two classes of these cyclin-dependent kinases inhibitors are recognized. **INK4A** family members inhibit D-type cyclins from associating with and activating CDK4 and CDK6. **CIP/KIP** family members are potent inhibitors of CDK2 kinases. p21 (p21^{CIP1}), described above, is a member of the CIP/KIP family.

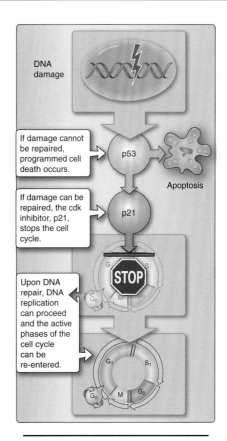

Figure 21.6
p53 control of cell fate following DNA damage.

Clinical Application 21.1: Human Papillomavirus, Cervical Cancer, and Tumor Suppressors

Human papillomavirus strains 16 and 18 are established etiological agents of cervical cancer. In cervical cells infected with these viral strains, viral protein E6 binds to p53 and viral protein E7 binds to RB. As a result of viral protein binding, both RB and p53 are inactivated and unregulated cell cycle progression and malignancy may result.

B. G_2 checkpoint

It is important for the integrity of the genome that nuclear division (mitosis) does not begin before DNA is completely duplicated during S phase. Therefore, the G_2 checkpoint, which occurs after S and before the initiation of mitosis, is also a critical regulatory point within the cell cycle. CDK inhibitors and phosphatases function at the G_2 checkpoint.

1. **Cyclin-dependent kinase inhibitor 1 (CDK1):** Cyclin-dependent kinase inhibitor 1 controls entry into mitosis. During G_1, S, and into G_2, CDK1 is phosphorylated on tyrosine residues, inhibiting its activity. In order for the cell to progress through G_2 and into M, those inhibitory phosphorylations must be removed from CDK1.

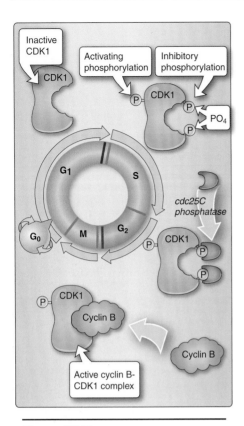

Figure 21.7
Progression from G_2 to M is controlled by cdc25C phosphatase and CDK1.

2. cdc25C phosphatase: cdc25C phosphatase is the enzyme that catalyzes the removal of inhibitory phosphorylations from CDK1 (Figure 21.7). Following its dephosphorylation, CDK1 can bind to cyclin B and the activated CDK1-cyclin B complex moves into the nucleus where it activates mitosis by phosphorylating key components of subcellular structures (e.g., microtubules). If the cell cycle must be suspended prior to chromosome segregation in mitosis, then cdc25C can be inactivated through actions of tumor suppressors **ATM** and **ATR** (see below).

IV. DNA DAMAGE AND CELL CYCLE CHECKPOINTS

DNA damage within cells may be caused in a variety of ways including replication errors, chemical exposure, oxidative insults, and cellular metabolism. The usual response to DNA damage is to halt the cell cycle in G_1 until DNA repair (Chapter 7) can be accomplished. As previously described, the tumor suppressor p53 responds to DNA damage by halting the cell in G_1. However, depending on the type of DNA damage, different cell cycle regulatory systems may be utilized, and other phases of the cell cycle may be halted. Additional tumor suppressor proteins can play a role in the control of checkpoints in cases of DNA damage.

A. ATM and ATR response to DNA damage

Tumor suppressors **ATM** (ataxia telangiectasia, mutated) and **ATR** (ATM and Rad3 related) are serine and threonine protein kinases that are important in the cellular response to DNA damage (Figure 21.8).

ATM is activated by ionizing radiation and is the primary mediator of the response to double-strand DNA breaks. It can induce cell cycle arrest at G_1/S, in S and at G_2/M transitions. ATR plays a role in arresting the cell cycle in response to UV-induced DNA damage and has a secondary role in the response to double-strand DNA breaks.

1. BRCA1: BRCA1, the protein product of the breast cancer susceptibility gene 1, plays a role in the repair of double-strand DNA breaks. It is involved in all phases of the cell cycle. The mechanistic details and other proteins involved remain to be elucidated.

V. ANTICANCER DRUGS AND THE CELL CYCLE

Both normal and tumor cells utilize the same cell cycle. But, normal and **neoplastic** (cancerous) tissue may differ in the total number of cells in active phases of the cell cycle. Some chemotherapeutic agents are effective only in actively cycling cells (Figure 21.9). These therapies are considered to be cell cycle–specific agents and are generally used for tumors with a high percentage of dividing cells. Normal, actively cycling cells are also damaged by such therapies. When tumors have a low percentage of dividing cells, then cell cycle–nonspecific agents can be used therapeutically.

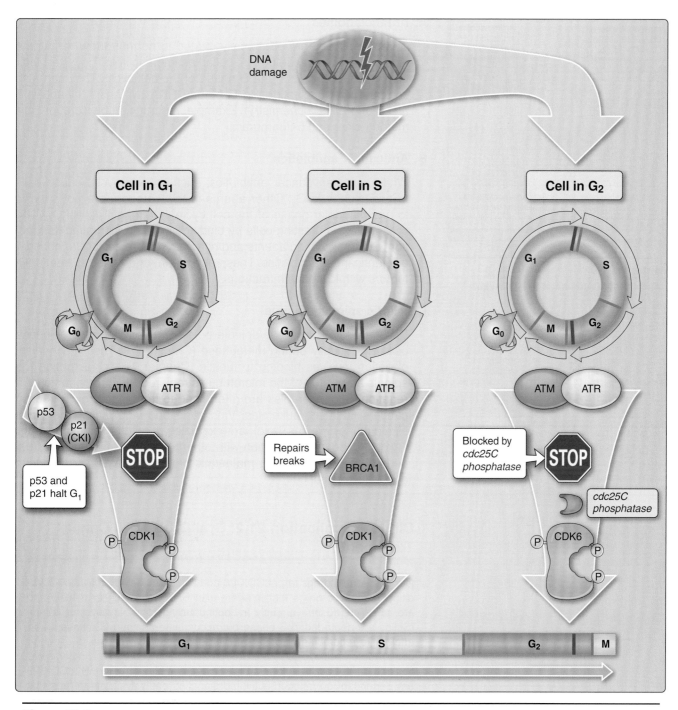

Figure 21.8
ATM/ATR in G_1 and G_2 checkpoint regulation.

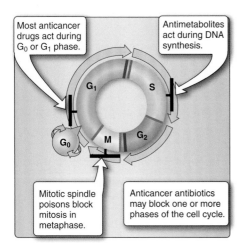

Figure 21.9
Anticancer drugs and the cell cycle.

A. Antimetabolites

Compounds that are structurally related to normal purine or pyrimidine nucleotide precursors are called antimetabolites. They exert their toxic effects on cells in the S phase of the cell cycle. They can also compete with nucleotides in DNA and RNA synthesis (Chapters 7 and 8 and *LIR Biochemistry*). Examples of drugs in this category are methotrexate and 5-fluorouracil.

B. Anticancer antibiotics

While some anticancer antibiotics, such as bleomycin, cause cells to accumulate in G_2. Other agents in this category are not specific to any particular phase of the cell cycle, but impact actively cycling cells more than resting cells by binding DNA and disrupting its function. Some alkylating agents and nitrosoureas are cell cycle–nonspecific anticancer antibiotics. These agents are often used to treat solid tumors with low growth fractions.

C. Mitotic spindle poisons

Drugs that act as mitotic spindle poisons inhibit mitotic or M-phase cells, specifically during metaphase (Chapter 20). Their mechanism of action involves binding to tubulin (Chapter 4) and disrupting the spindle apparatus of the microtubules required for chromosome segregation. Such therapies are often used to treat high growth fraction cancers such as leukemias. Vincristine and vinblastine (the Vinca alkaloids), as well as Taxol, are examples of mitotic spindle poisons. Taxol is used in combination with other chemotherapy drugs to treat certain cancers, including metastatic breast cancer, ovarian cancer, and testicular cancer.

Clinical Application 21.2: Oral Drugs and Cancer Treatment

Newly available drugs target CDK4/6 and can be ingested orally, unlike traditional cytotoxic drugs that interfere with DNA replication or mitosis and are administered intravenously. In continuously cycling cancer cells, D-type cyclins are degraded in S phase but accumulate again in G_2 phase, in an attempt to recombine with CDK4/6 in subsequent G_1 phases to allow for uninterrupted cell cycles. The CDK4/6 inhibitors impair cyclin D recombination with CDK4/6. Since the stability and assembly of cyclins with CDK4/6 are dependent on mitogen-activated signaling pathways, using drugs that inhibit the MAP kinase pathway along with drug inhibitors of cyclin-CDK4/6 interactions appears to produce synergistic effects. This combination of drugs allows the cell to arrest in G_1 and exit from the cell cycle into a quiescent (G_0) state. Monotherapy with CDK4/6 inhibitors has not yet proven to be effective in controlling cancers, but their use in combination with these MAP kinase inhibitors appears to hold promise.

Chapter Summary

- Cyclins and CDKs control cell cycle progression.

- Specific cyclins are made and degraded during particular points in the cell cycle.

- CDKs are enzymatically active during a narrow window of the cell cycle and are important for driving the cell cycle forward.

- Cells are stimulated to enter the cell cycle by the action of growth factors, which directly activate a specific cyclin belonging to the cyclin D family of G_1 cyclins.

- Tumor suppressor proteins inhibit the cell cycle. Mutated tumor suppressors allow unregulated cell cycle progression that may result in malignancy.

- Checkpoint regulation is a safety measure to prevent accumulated DNA damage.

- RB protein controls movement into the S phase by inhibiting S phase–specific transcription factors. RB is active in its underphosphorylated form and inactive in its hyperphosphorylated form.

- p53 guards the genome from damage. In the event of DNA damage, p53 can induce the synthesis of a CKI.

- CDK1 controls the G_2M transition and is activated by the phosphatase action of cdc25C.

- ATM and ATR are kinases that sense and respond to specific types of DNA damage.

- External and internal factors contribute to different types of DNA damage.

- Anticancer drugs may have cell cycle–specific or cell cycle–nonspecific actions.

- Antimetabolites inhibit S-phase cells, while anticancer antibiotics may cause accumulation of G_2-phase cells or act without regard to cell cycle phase. Mitotic spindle poisons disrupt spindle formation and affect cells in mitosis.

Study Questions

Choose the ONE best answer.

21.1 Which of the following types of cells are senescent and permanently in G_0?
- A. Embryonic stem cells
- B. Hematopoietic cells
- C. Hepatocytes
- D. Intestinal epithelial cells
- E. Neurons

Correct answer = E. Neurons that have completed mitosis do not reenter the cell cycle. Embryonic stem cells (Chapter 1) have an enormous capacity for division and differentiation. Hematopoietic cells, which arise in bone marrow, continue to divide in order to replace lost blood cells and/or to create cells necessary for an immune response. Hepatocytes, functional cells in the liver, retain the ability to divide, allowing the liver to have a great capacity for regeneration. Intestinal epithelial cells divide quickly as they must be able to be replaced following injury that occurs as a normal consequence of their function and location.

21.2 In a cell with only mutant versions of RB, which of the following actions will be inhibited?
- A. Activation of DNA synthesis
- B. Arrest of the cell cycle in G_1
- C. Binding of cyclins to CDKs
- D. Completion of nuclear division during mitosis
- E. Removal of inhibitory phosphorylations from CDKs

Correct answer = B. Arrest of the cell cycle in G_1. Mutant RB does not halt the cell cycle in G_1. Instead, it will allow unregulated passage out of G1. Activation of DNA synthesis will occur. Cyclins will bind to CDKs during cell cycle progression and inhibitory phosphorylations will be removed (by cdc25C phosphatase) from CDK1, as is required for progression through G_2. RB has no role in mitosis.

21.3 If a normal, actively cycling cell receives damage to its DNA while it is in G_1 phase, which of the following will function to arrest the cell cycle?
A. cdc25C phosphatase
B. Cyclin D
C. CDK2
D. E2F
E. p21^{CIP1}

Correct answer = E. p21^{CIP1} is a CKI that is activated by p53. Its role is to halt cell cycle progression in order to allow for DNA repair to take place. cdc25C phosphatase dephosphorylates CDK1, which controls entry into mitosis. Cyclin D activates CDK4 or CDK6 to allow for progression through G1 to S. CDK2 is activated by cyclin E or A to initiate DNA synthesis in early S phase. E2F is a transcription factor critical for the G_1/S transition.

21.4 In a normal, cycling cell in S phase, which of the following proteins will be active?
A. BRCA1
B. CDK2
C. p21
D. p53
E. RB

Correct answer = B. Of the choices listed, only CDK2 is active during S phase. BRCA1, the breast cancer susceptibility gene, plays a role in the repair of double-strand DNA breaks and does not function in a normal, cycling cell. p21 is a CDK inhibitor induced by p53 in response to DNA damage. Normal RB halts cells in G1 when it is not appropriate for cells to continue through the remainder of the cell cycle.

21.5 Replication errors are detected in a normal cell that has just completed S phase. The usual cellular response to this DNA damage will include
A. Binding of cyclin D to CDK2.
B. Halting the cell cycle at the restriction point.
C. Inactivation of cdc25C phosphatase.
D. Inhibition of purine nucleotide biosynthesis.
E. RB binding to transcription factor E2F.

Correct answer = C. Upon inactivation of cdc25C phosphatase, the cell described must halt the cell cycle in G_2. Control at the G_2 checkpoint involves inactivation of cdc25C phosphatase, preventing dephosphorylation of CDK1 and halting the cell cycle to allow for DNA repair. Binding of cyclin D to CDK2 allows for progression within G_1. The restriction point is within G_1, and this cell is in G_2. Inhibition of purine nucleotide biosynthesis is the action of some cell cycle–specific cancer chemotherapy drugs. RB binding to E2F prevents entry into S phase. This cell has completed S phase.

21.6 A biopsy of a liver mass found in a 79-year-old female reveals malignant cells that contain a mutant tumor suppressor protein. Which of the following proteins is most likely present in mutated form in this liver tumor?
A. CDK4
B. Cyclin B
C. Cyclin D1
D. p21^{CIP1}
E. p53

Correct answer = E. Over 50% of cancer cells have mutant p53 tumor suppressor genes. p21^{CIP1} is the CKI induced by functional p53 to arrest the cell cycle in G_1. The other proteins, cyclins and CDKs, function to activate the cell cycle.

21.7 A 72-year-old male, recently diagnosed with bladder cancer, is undergoing chemotherapy treatment with methotrexate. Two weeks following his first treatment, the patient presents with weakness, hair loss, and oral ulcers. Which of the following best explains the mechanism underlying these signs and symptoms?
A. Actively cycling normal cells are destroyed by the drug.
B. Cell cycle–nonspecific effects of the drug damage normal cells.
C. Continued tumor growth causes damage to normal body cells.
D. Present signs and symptoms likely result from an undiagnosed pathology.
E. Tumor cell lysis results in damage to normal body cells.

Correct answer = A. Normal cycling cells are destroyed by methotrexate, a cell cycle–specific antimetabolite that exerts its effects on S-phase cells. This drug can damage any cell in the body that arrives at S phase. The patient's weakness most likely results from anemia owing to inhibition of production of red blood cells (erythrocytes). Methotrexate is not a cell cycle–nonspecific drug. Such drugs would not likely damage actively cycling cells to the extent seen in this patient. Continued tumor growth and tumor cell lysis may damage other body cells, but nearby cells may be most affected. In this patient's situation, the affected cell types causing the signs and symptoms are all actively cycling cells. While an undiagnosed pathology may exist, the most likely explanation for the signs and symptoms is that normal cycling cells have been damaged by the S phase–specific actions of methotrexate.

Abnormal Cell Growth

22

I. OVERVIEW

Cells are often lost through death by apoptosis or necrosis, by sloughing or shedding, or by injury. New cells normally replace cells at the same rate they are lost, in a highly regulated state of balance known as **homeostasis**. If normal cellular regulatory mechanisms malfunction, unregulated and unchecked cell division may result, and **cancer** results.

Protooncogenes regulate or produce proteins that coordinate normal cell growth and development. Mutations that alter protooncogenes can convert them from regulatory genes into cancer-causing **oncogenes**. In addition, mutations that create a loss of function in tumor suppressor genes may also induce cancer.

Most genetic changes that occur during transformation of normal cells to cancer cells (carcinogenesis) are somatic mutations. Each time a cell divides, there is a chance of a somatic mutation; therefore, there is always a low background risk for cancer. A far more prevalent cause of cancer is environmental exposure.

II. GENES AND CANCER

Cell division is controlled by a number of cellular proteins. Because these proteins are the products of genes, genetic mutation may result in unregulated cellular proliferation. Protooncogenes normally promote cell cycle progression, and tumor suppressor genes normally function to control progression through the cell cycle. Mutations in protooncogenes and in tumor suppressor genes can both lead to cancer.

A. Protooncogenes and oncogenes

Protooncogenes are genes whose protein products control cell growth and differentiation. These genes can undergo mutations into **oncogenes** that cause qualitative and quantitative changes in their protein products. Our knowledge of protooncogenes stems from molecular genetic studies on the defective gene product. Protooncogenes have been identified in the various signal transduction cascades that control cell growth, proliferation, and differentiation. As normal regulatory elements, protooncogenes function in a wide variety of cellular pathways (Figure 22.1).

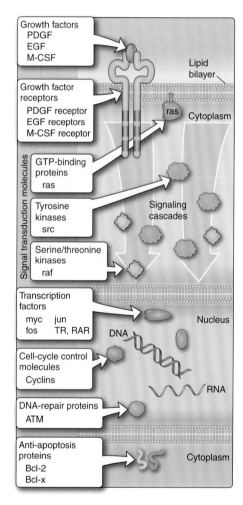

Figure 22.1
Protooncogenes and their roles in regulating growth.

Mutations can occur in protooncogenes involved in any of the steps involved in regulating cell growth and differentiation. When such mutations accumulate within a particular cell type, the progressive deregulation of growth eventually produces a cell whose progeny forms a tumor. Point mutations, insertion mutations, gene amplification, chromosomal translocation and/or changes in expression of the oncoprotein can all result in deregulated activity of these genes (Figure 22.2).

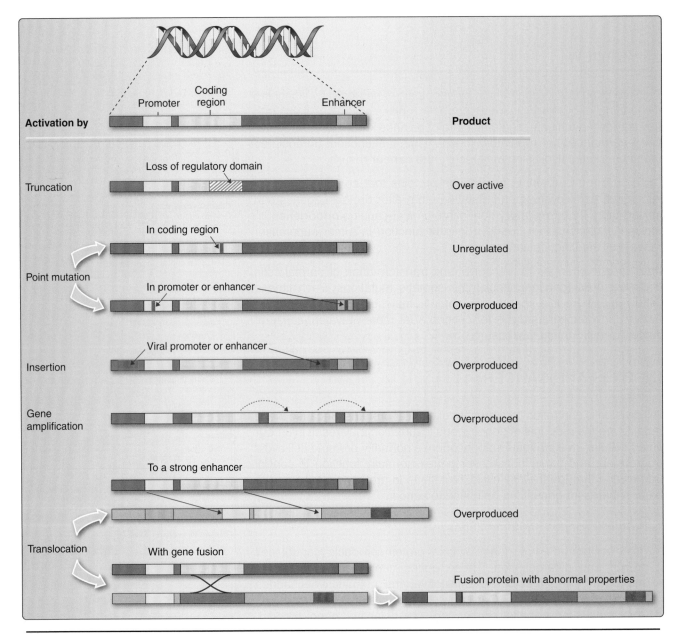

Figure 22.2
Mechanism of conversion of protooncogenes to oncogenes.

B. Tumor suppressor genes

Tumor suppressor genes are important for maintaining normal cell growth control by curtailing unregulated progression through the cell cycle. Situations that diminish tumor suppressor gene function may lead to neoplastic changes.

Mutations in tumor suppressor genes predisposes cells to cancer. Protein products of tumor suppressor genes normally repress cell growth and division. Therefore, **loss of function**, through mutations or other alterations, can lead to malignant transformation by removing the restraints that normally regulate cell growth.

1. **Retinoblastoma:** The retinoblastoma gene, RB, was cloned in 1987 and was the first tumor suppressor gene cloned. RB functions to prevent excessive cell growth by inhibiting cells in G1 (see Chapter 21). Inactivation of RB via phosphorylation inactivates it and allows the cell to progress into S. Mutant RB encodes a dysfunctional protein that permits unregulated progression out of G_1.

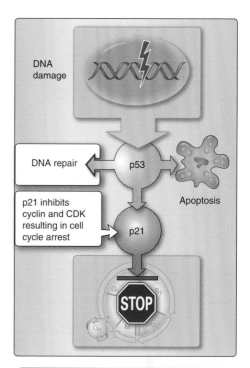

Figure 22.3
p53—guarding the genome.

Clinical Application 22.1: Retinoblastoma Cancer

Retinoblastoma is an embryonic malignant neoplasm that begins in the retina of the eyes. The genetic locus responsible for a predisposition to retinoblastoma is located within the q14 band of chromosome 13. It is estimated that 40% of cases of retinoblastoma cancer are hereditary and 60% are sporadic (nonhereditary). The hereditary form is caused by mutations transmitted in the germ line of one of the parents, and therefore, affected persons begin life with one mutant copy of RB in every cell within their body. Mutation of the second, wild-type or normal copy of RB is likely to happen, causing malignancy, initially in the retinas. Patients with the germ line type of retinoblastoma have a markedly increased frequency of second malignant tumors—the most common being osteosarcomas. Individuals with the sporadic form of retinoblastoma acquired RB mutations early in life, but did not inherit them.

---

2. **p53—guardian of the genome:** The most frequently inactivated tumor suppressor gene is the **p53 gene,** which encodes a protein with a 53 kilodalton molecular mass, or **p53,** which is often implicated in cancer development. More than half of human cancers show p53 mutations (Chapter 21). Loss of p53 function can contribute to genomic instability within cells (Figure 22.3). Functional p53 is important in preventing cancer because of its unique functional capabilities.

   p53

   • regulates gene expression and controls several key genes involved in growth regulation

   • facilitates DNA repair. When DNA damage is encountered, p53 senses the damage and causes G_1 arrest of the cells, until the damage is repaired.

   • activates apoptosis of damaged cells. When damage to DNA within cells is beyond repair, p53 functions to trigger apoptosis in these cells.

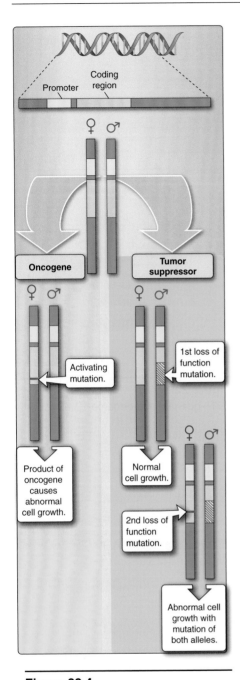

Figure 22.4
Oncogenes are dominant in action at the cell level, and tumor suppressor genes are recessive.

3. **Nature of oncogenes and tumor suppressor genes:** Some genetic mutations confer a growth advantage to the cell that contains it, allowing selective growth of these cells. Therefore, when protooncogenes undergo mutations, they are "activated" to oncogenes (Figure 22.4). Because these genes normally regulate growth, mutations in them often favor the unregulated growth of cancer.

Generally, tumor suppressor genes are "inactivated" by mutations and deletions, resulting in loss of function of the protein and in unregulated cell growth. Both copies of the tumor suppressor genes have to be mutated or lost for loss of growth control; therefore, these genes act recessively at the cellular level. Oncogenes, on the other hand, are dominant in action requiring mutation of only one copy of a protooncogene (Figure 22.4).

Clinical Application 22.2: MicroRNAs as Oncogenes and Tumor Suppressors

As a class, **microRNAs (miRNAs)** regulate gene expression by controlling the levels of target RNA posttranscriptionally. They are encoded within the noncoding and intron regions of different genes and are transcribed into RNA, but not translated into protein. These single-stranded RNA molecules are approximately 21 to 23 nucleotides in length and are processed from primary transcripts known as *pri-miRNA* into short stem-loop structures, *pre-miRNA*, and finally to functional miRNA. Mature miRNA molecules are partially complementary to one or more messenger RNA (mRNA) molecules and function to down-regulate gene expression. Current knowledge suggests the presence of about 1,000 microRNA genes that appear to target over 60% of the genes in the mammalian genome.

miRNAs affect the expression of critical proteins within the cell, such as cytokines, growth factors, transcription factors, etc. The expression profiles of miRNAs are frequently altered in tumors. When miRNA's targets are oncogenes, their loss of function results in increased target gene expression. Conversely, overexpression of certain miRNAs may decrease the levels of protein products of target tumor suppressor genes. Therefore, miRNAs behave as oncogenes and tumor suppressor genes. New insights into miRNAs function may also advance the diagnosis and treatment of cancer.

III. MOLECULAR BASIS OF CANCER

Normal cells respond to a complex set of biochemical signals, which allow them to develop, grow, differentiate, or die. Cancer results when any cell is freed from these types of restrictions and the resultant abnormal progeny of cells are allowed to proliferate.

Development of cancer is a stepwise process. Often, several genetic alterations must occur at specific sites before malignant transformation is seen in adult cancers. Cancers of childhood appear to require fewer mutations before manifestation of overt cancer. Rare inherited mutations present in virtually all somatic cells of the body can predispose individuals to cancer at one or more sites.

A. Cancer genesis—a multistep process

Mutations in key genes must accumulate over time in order to create a progeny of cells that have lost growth control. Each individual mutation contributes in some way to eventually producing the malignant state. The accumulation of these mutations spans several years and explains why cancers take a long time to develop in humans (Figure 22.5). Both exogenous (environmental insults) and endogenous processes (carcinogenic products generated by cellular reactions) may damage DNA. DNA damage that goes unrepaired may lead to mutations during mitosis. Increased errors during DNA replication or a decreased efficiency of DNA repair may favor increased frequency of genetic mutations. Cells also become cancerous when mutations occur in protooncogenes and tumor suppressor genes.

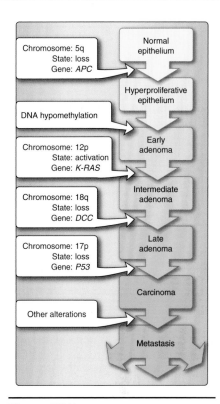

Figure 22.5
Colon cancer progression.

Clinical Application 22.3: Driver and Passenger Mutations

Since it is now possible to easily determine the sequence of all coding genes within several cancers, it has allowed for the discovery of the specific mutations in these cancers. Such studies have helped determine that not all mutations are responsible for initiating the tumor. In fact, a small number of genes, when mutated, confer a growth advantage to cancer cells, suggesting the existence of "driver genes." The remaining genes that undergo mutations in cancers are considered "passengers" as their mutations might have occurred coincidently during the progression of the early cancerous lesions. These findings have led to the suggestion that the set of driver genes identified for specific cancers can be exploited in targeted cancer therapies. To obtain clinically useful responses, these driver gene mutations should be targeted as they probably existed in both the primary and metastatic lesions.

B. Theories of cancer

It has been long known that cancer cells are genetically unstable. Only in the last two decades has it been realized that specific genes are responsible for this instability. Nearly 200 oncogenes and 170 tumor suppressor genes have been identified. Additional genes that aid in breaking down basement membranes of cells and allow for their movement are also known to be important in oncogenesis. Despite the large number of possible genetic permutations, certain combinations of these mutant genes were found in distinct cancers and also in different types of cancer from the same tissue. From these observed patterns came several distinct theories of cancer formation.

1. **Clonal evolution model:** This model was proposed in the 1970s to explain how cancers evolve. According to this model, initial damage (a genetic mutation) occurs in a single cell, giving it a selective growth advantage and time to outnumber neighboring cells. Within this clonal population, a single cell may acquire a second mutation, providing an additional growth advantage and allowing it to expand and become the predominant cell type. Repeated cycles followed by clonal expansion eventually lead to a fully developed malignant tumor. Accumulated mutations within key genes trigger a single transformed cell to eventually develop into a malignant tumor (Figure 22.6).

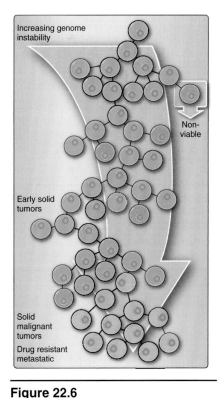

Figure 22.6
Theory of clonal evolution.

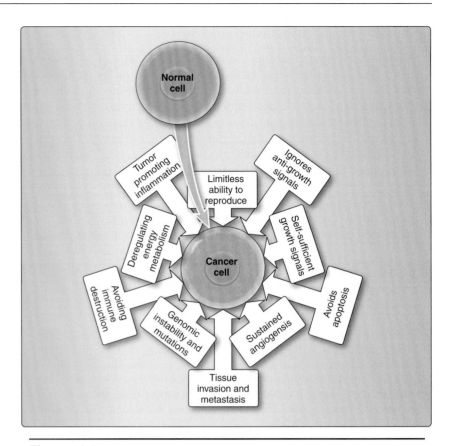

Figure 22.7
Hallmarks of cancer.

2. **Hallmarks of cancer model:** The number of genes identified in cancers is constantly growing, and the complexity of these observations has been recently streamlined into a series of genetic and cellular principles that may govern the formation of most, if not all, types of human cancers (Figure 22.7). According to this model, oncogenesis requires cells to

   - acquire self-sufficiency of growth signals
   - become insensitive to growth inhibitory signals
   - evade apoptosis
   - acquire limitless replicative potential
   - sustain angiogenesis
   - acquire capabilities to invade tissues and metastasize
   - create genome instability
   - promote inflammation
   - avoid immune destruction
   - reprogram energy metabolism

   In the hallmarks model, the type of genetic insult may vary with different cancers. All cancers, however, should acquire damage to these different classes of genes until a cell loses a critical number of growth control mechanisms and initiates a tumor.

3. **Stem cell theory of cancer:** This theory accounts for the observations that tumors contain cancer stem cells with indefinite proliferative potential similar to adult stem cells. Because hematopoietic stem cell lineages have been well characterized, most of our evidence for cancer stem cells has come from leukemias. Cancer stem cells are thought to be self-renewing and responsible for all components of a heterogeneous tumor. These tumor-initiating cells tend to be drug resistant and to express markers typical of stem cells. The cancer stem cell model is also consistent with some clinical observations, in which standard chemotherapy has not been successful in destroying all tumor cells and some cells remain viable. Despite the small number of cancer stem cells, according to this theory, they may be responsible for tumor recurrence years after "successful" treatment (Figure 22.8). Several genes have been identified that can confer self-renewal properties to committed progenitors and mediate their neoplastic transformation.

C. Tumor progression

Cancer cells gain metastatic abilities as they evolve. Among these are genes whose products allow the breakdown of tissue structure and invade the basement membrane, allowing cells to migrate to other sites. In addition, as tumors accumulate in cellular mass, it is critical that they induce the growth of blood vessels, or **angiogenesis**, to supply the growing tumor with adequate nutrition and oxygen for its continued growth and survival.

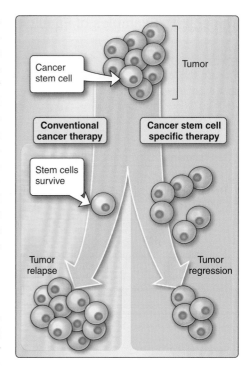

Figure 22.8
Stem cell theory of cancer.

Clinical Application 22.4: Angiogenesis and Tumor Progression

As cancers progress, they accumulate in mass. To obtain the necessary nutrition and oxygen for their continued growth and survival, it is critical for tumors to have adequate blood supply. Cancer cells create new blood supply to sustain their growth by several mechanisms. Angiogenesis, or **neovascularization**, can be both activated and inhibited. Under normal physiologic conditions, angiogenic inhibitors predominate, blocking the growth of new blood vessels. When there is a need for new vasculature, activators increase in number and inhibitors decrease. Tumors release two proangiogenic factors, which are important for sustaining tumor growth: **vascular endothelial growth factor (VEGF)** and **basic fibroblast growth factor (bFGF)**. Several angiogenesis inhibitors and antibodies to proangiogenic growth factors are currently in clinical trials to inhibit tumor progression. One mechanism for tumor neovascularization involves mutation of *p53* gene. Typically, wild type p53 regulates the expression of an angiogenesis inhibitor, **thrombospondin**. Mutations in p53 facilitate neovascularization due to the lack of production of thrombospondin.

IV. INHERITED MUTATIONS AND CANCER

The number of individuals with an inherited predisposition to cancer is low when compared to the total number of human cancers. But, for the individual carrying a mutation in a cancer-causing gene, the risk for cancer is several-fold higher of developing the cancer. Since such mutations

Table 22.1: Examples of Familial Cancer Syndromes

Syndrome	Primary Tumor	Associated Conditions	Gene	Function of Gene Product
Familial breast cancer	Breast cancer	Ovarian cancer	*BRCA1*	Repair of double strand DNA breaks
	Breast cancer	Ovarian cancer Pancreatic cancer Melanoma	*BRCA2*	Repair of double strand DNA breaks
Li-Fraumeni	Sarcomas Breast cancer	Leukemias Brain tumors	*P53*	Transcription factor Apoptosis cell cycle arrest
Hereditary nonpolyposis colon cancer	Colorectal cancer	Endometrial Ovarian Bladder Glioblastoma	*MSH2 MLH1*	Repair of DNA mismatches Maintains DNA stability
Familial adenomatous polyposis	Colorectal cancer	Duodenal Gastric tumors	*APC*	Regulation of β catenin levels Cell adhesion
Familial retinoblastoma	Retinoblastoma	Osteosarcoma	*RB*	Cell cycle regulator Transcription factor

are inherited through the germ line, they are therefore present in every cell of the body.

A large percentage of genes that are mutated in familial cancers are in tumor suppressor genes. Some examples are shown in Table 22.1. Mutation in protooncogenes during development may not be compatible with life. This highlights the importance of orderly growth during embryogenesis to produce a viable fetus, which is better facilitated with the loss of a copy of a tumor suppressor gene than in the presence of an oncogene.

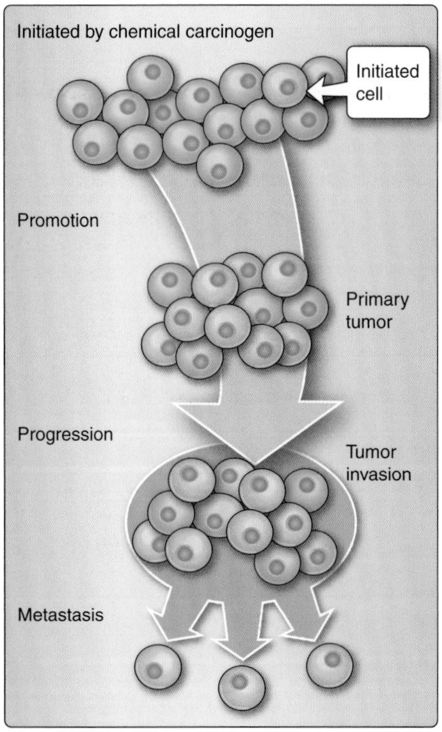

Figure 22.9
Chemical carcinogenesis.

V. MUTATIONS IN DRUG-METABOLIZING ENZYMES AND CANCER SUSCEPTIBILITY

While inheritance of cancer-causing genes will increase the cancer risk, its occurrence in the population is low. On the other hand, the carcinogen-metabolizing enzymes exist in different forms within the population at a high rate and will increase the cancer risk in some individuals carrying the form that allows activation of certain carcinogens.

Environmental chemicals may be classified as either **genotoxic** chemicals that interact with DNA, causing mutation in critical genes, or **nongenotoxic,** whose mechanisms differ depending upon the nature of the chemical compound. Chemical carcinogenesis is a multistep process (Figure 22.9). Chemicals that have no appreciable carcinogenic potential of their own but greatly enhance tumor development when exposed to them for long periods of time mediate tumor promotion. In terms of lifestyle, exogenous hormones, high-fat diet, alcohol, etc. are known to promote cancer and therefore can be an important determinant of cancer risk.

While genetic predisposition, ethnicity, age, gender, and, to some extent, health and nutritional impairment are cancer susceptibility factors, recent studies are showing polymorphisms in certain drug-metabolizing enzymes to be associated with this interindividual variation (Figure 22.10). Variations in the expression or form of the drug-metabolizing genes, such as **cytochrome P450, glutathione transferase,** and ***N*-acetyl transferase genes,** strongly influence individual biologic response to carcinogens.

Clinical Application 22.5: Mutations in P53 Mirror Etiological Agent in Human Carcinogenesis

The p53 tumor suppressor gene is mutated in over 50% of lung, breast, colon, and other common tumors; the mutational spectrum varies by cancer type and environmental exposure, providing clues to the specific risk factors involved.

Mutations in the specific codon of the p53 gene occur in lung, head, and neck cancers when benzopyrene adducts (cigarette smoke, environmental carcinogens) bind to DNA. In geographic areas in which aflatoxins and hepatitis B are risk factors for liver tumors, specific p53 mutations (codon 249, AGG to AGT) are seen to occur in these tumors.

Alterations seen in the ***P53*** gene in squamous and basal cell carcinomas of skin are hallmarks of exposure to ultraviolet light. Cervical tumors are due to human papillomavirus (HPV) infection. In HPV-positive tumors, p53 remains wild type, but binding to HPV causes its rapid degradation. p53 is an example of how several of the capabilities of cancer are realized with its loss and underscores its importance in the prevention of cancer.

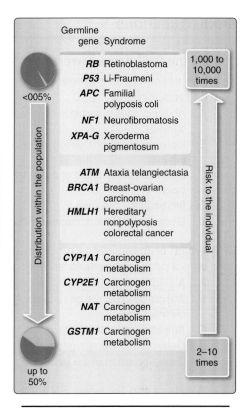

Figure 22.10
Drug-metabolizing enzymes and cancer risk.

Chapter Summary

- Cancer is a multistep process.
- There are several characteristic features that a cell must attain before it undergoes neoplastic transformation.
- Protooncogenes are normal counterparts of oncogenes and usually have a role in growth regulation. Protooncogenes undergo mutations that cause them to be overactive or function without regulation.
- Tumor suppressor genes normally restrain growth. Tumor suppressor genes lose function when mutated.
- Oncogenes are dominant in action, and tumor suppressor genes are recessive.
- Mutations in DNA repair genes can cause cancer.
- Germ line mutations in cancer-causing genes occur infrequently. Inherited predispositions to cancer account for 5% to 10% of all human cancers.
- Lifestyle factors influence cancer risk in the general population.
- Polymorphism in drug-metabolizing enzymes explains cancer susceptibility in the general population.
- Mutations in p53 gene are the most common cancer-associated mutations, and its function underscores its importance in the prevention of cancer.

Study Questions

Choose the ONE best answer.

22.1 A mutation in p53 that causes a loss of its function can result in a(n)

A. Ability of cells to arrest in G_1 phase after DNA damage.

B. Increase in the production of an angiogenesis inhibitor.

C. Decrease in induction of apoptosis of damaged cells.

D. Increase in DNA repair.

E. Decrease in DNA damage within cells.

> Correct answer = C. A loss of p53 function increases the survival of damaged cells, since p53 is no longer able to sense the presence of DNA damage in them. Functional p53 can arrest the cell cycle in the G_1 phase, transcriptionally activate the production of an angiogenesis inhibitor, and allow repair of damaged DNA. The presence of mutant p53 will increase DNA damage within cells.

22.2 Which of the following mechanisms cannot activate a protooncogene to an oncogene?

A. A single point mutation in the gene

B. Translocation of the gene to a site on a different chromosome

C. Amplification of the gene

D. Deletion of the entire gene

E. Overexpression of the gene

> Correct answer = D. Deletion of the gene will result in complete loss of expression. All the other modifications have the potential to produce an aberrant oncogenic protein.

22.3 Which of the following will increase the risk of neoplastic transformation?

A. Increase in the activity of DNA repair enzymes

B. Decrease in the rate of mutations of protooncogenes

C. Decrease in the activity of tumor suppressor genes

D. Increase in the activity of carcinogen-metabolizing enzymes

E. Decrease in cell cycle activity

> Correct answer = C. Decreased tumor suppressor gene activity allows DNA damage to accumulate and allows cell cycle to be unregulated, all of which will increase the risk of neoplastic transformation. An increase in DNA repair is beneficial to the overall health of the cell. Mutations in protooncogenes increase cancer occurrence, so a decrease will prevent neoplastic transformation. An increase in carcinogen-metabolizing activity will help remove potential carcinogens from the body. A decrease in the cell cycle will prevent abnormal growth.

22.4 Cancer risk in which of the following individuals is the highest based on the knowledge of their health histories? A 24-year-old

A. Diabetic male

B. Obese male with a family history of cardiovascular disease

C. Premenopausal woman taking exogenous hormones

D. Underweight female on a perpetual diet

E. Woman with a family history of breast cancer

> Correct answer = E. A woman with a family history of breast cancer is likely to carry mutations in the predisposing gene. The mutant gene is now present in every cell of her body. Having a mutant gene makes the cell susceptible to further mutations. Given the current knowledge, her risk is the highest for cancer occurrence, but it does not mean that she will definitely have cancer. The presence of diabetes or cardiovascular disease does not raise the risk of cancer. Being underweight does not increase cancer risk. Exogenous estrogen is known to slightly increase cancer risk.

22.5 Which of the following proteins has the potential to inhibit angiogenesis?

A. Adenomatous polyposis coli (APC)

B. Telomerase

C. Thrombospondin

D. Retinoblastoma

E. N-acetyl transferase

> Correct answer = C. Thrombospondin is an angiogenesis inhibitor. APC mutations predispose individuals to polyps in the colon. Telomerase activation is required for the tumor's "immortalization." Retinoblastoma is a tumor suppressor gene that controls the G_1 transition. N-acetyl transferase is a drug-metabolizing enzyme whose polymorphic forms may influence cancer risk in the population.

Cell Death

23

I. OVERVIEW

All cells eventually die by either necrosis or apoptosis. Necrosis is a passive, pathological process induced by cellular injury or accidental means and often involves the simultaneous death of cells in groups (Figure 23.1). Necrotic cells have ruptured cell membranes allowing the cytoplasm and organelles to spill into the surrounding tissue fluids, often inducing an inflammatory response. In contrast, apoptosis is an active, normal, physiological process that removes individual cells without damaging neighboring cells or inducing inflammation. Cells undergoing apoptosis have a characteristic "blebbed" appearance of their membranes. Apoptosis is as fundamental to cellular and tissue physiology as cell division and differentiation. Disturbance in pathways that regulate apoptosis may result in cancers, autoimmune diseases, and neurodegenerative disorders.

II. NECROSIS

Necrosis is a passive, pathological process induced by acute injury or disease. A group of cells in a localized region of a tissue generally undergoes necrosis at the same time after experiencing an insult. Cells that die by necrosis increase in volume and lyse (burst), releasing their intracellular contents. Mitochondria and other intracellular components are released, often inducing a potentially damaging **inflammatory response**. The necrotic process is completed within several days.

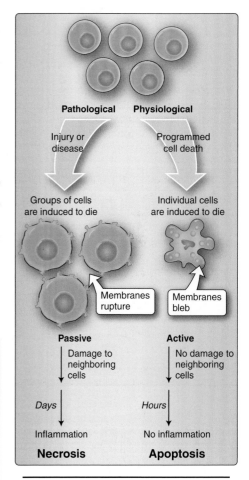

Figure 23.1
Cell death by necrosis and apoptosis.

Clinical Application 23.1: Necrosis and Serum Enzymes

Because intracellular contents, including enzymes, are released from necrotic cells, measurement of enzymes in serum samples prepared from a patient's blood is often done to aid in diagnosis and to help to determine a prognosis. For example, most cells contain lactate dehydrogenase (LDH), an enzyme that all cells use to generate ATP from glucose. When cells from any tissue die by necrosis, LDH will appear in the blood. In fact, LDH is often used as a general marker of necrotic cell death.

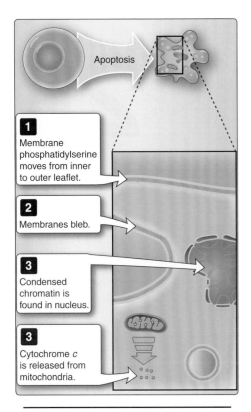

Figure 23.2
Cellular changes during apoptosis.

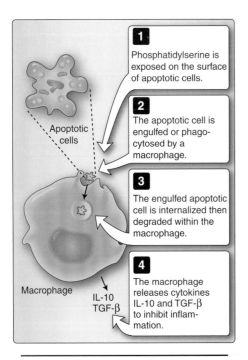

Figure 23.3
Apoptotic cell removal via phagocytosis.

III. APOPTOSIS

Cells deprived of survival factors activate an intracellular suicide program and die by a process of programmed cell death called **apoptosis** (pronounced ă-pŏp-tō'sĭs or as apo tō' sis). The requirement of a cell to receive signals for survival helps to ensure that cells continue to live only when and where they are needed.

Cells undergoing apoptosis shrink in size but do not lyse. Their plasma membrane remains intact, but portions of the membrane eventually bud off, or **bleb,** and lose their asymmetry and ability to attach to neighboring cells in a tissue. The membrane phospholipid **phosphatidylserine,** which is normally present on the inner membrane leaflet oriented toward the cytosol, inverts and becomes exposed on the cell's surface. In an active, ATP-requiring process, mitochondria of apoptotic cells release **cytochrome *c*** but remain within the membrane blebs (Figure 23.2). Chromatin of apoptotic cells segments and condenses.

Apoptotic cells are **engulfed by phagocytic cells**, macrophages and dendritic cells, which bind to the phosphatidylserine on the membrane surface (Figure 23.3). A macrophage internalizes and then degrades an apoptotic cell, reducing the risk of inflammation from the cell death. Phagocytic cells also release cytokines including interleukin-10 (IL-10) and transforming growth factor-β (TGF-β) that inhibit inflammation. Therefore, there is no extensive damage done to neighboring cells in a tissue when a nearby resident cell undergoes apoptosis. Apoptosis is completed within a few hours.

A. Biologic significance

While necrosis is a traumatic process resulting in widespread cell death, tissue damage, and inflammation, apoptosis has the advantage of eliminating individual cells whose survival would be harmful to the organism or whose elimination is critical for normal development or function.

1. **Removal of damaged cells:** Elimination of damaged cells is an important function of apoptosis. When a cell is damaged beyond repair, infected with a virus, or experiencing starvation or the effects of ionizing radiation or toxins, actions of the tumor suppressor protein **p53** (a product of the p53 gene, see also Chapter 21) halt the cell cycle and stimulate apoptosis (Figure 23.4). The normal (wild type) p53 binds to a p53-responsive element within the gene promoter of the proapoptotic protein **Bax**, triggering programmed cell death. Removal of individual cells by apoptosis saves nutrients needed by other cells and can also halt the spread of a viral infection to other cells. But mutant forms of p53 can neither halt the cell cycle nor initiate apoptosis. Therefore, abnormal cells expressing mutant p53 can continue to divide and fail to undergo apoptosis, despite the fact that their survival damages the organism.

2. **During development:** Apoptosis is used during development of the embryo. Extensive cell division and differentiation during this period often result in an excess number of cells that must be removed in order for normal development to proceed and for normal function to occur. In a developing vertebrate nervous system,

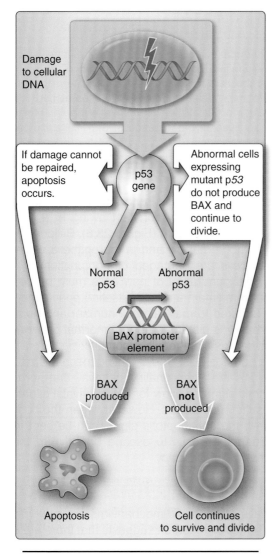

Figure 23.4
Apoptosis in response to DNA damage; role of p53 in apoptosis.

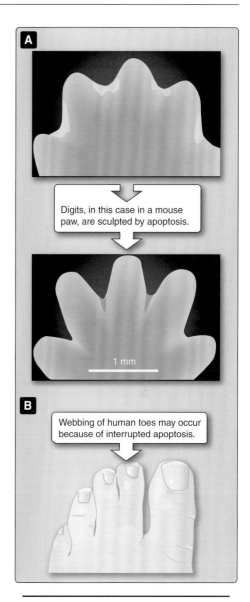

Figure 23.5
Sculpting by apoptosis.

more than half of the nerve cells generated undergo programmed cell death soon after they are formed.

Selective apoptosis "sculpts" the developing tissues. For example, apoptotic death of cells between developing digits must occur for formation of individual fingers and toes (Figure 23.5A). Incomplete apoptosis can result in abnormal structures (Figure 23.5B). Development of a healthy and mature adaptive immune system also requires apoptosis. Negative selection in the thymus, the process by which autoreactive T cells are eliminated from the repertoire of cells, likewise occurs via apoptosis (see also *LIR Immunology*, p. 114).

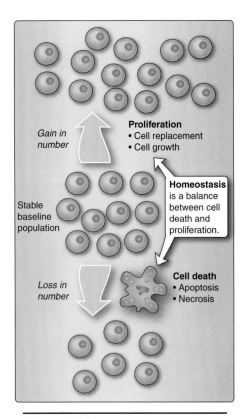

Figure 23.6
Homeostatic balance is maintained by a
balance of cell growth and cell loss.

3. **For tissue homeostasis:** In normal, healthy adults, the number
   of cells is kept relatively constant owing to a balance between
   cell division and cell death (Figure 23.6). Billions of cells die each
   hour in the bone marrow and in the epithelia of healthy individuals. When cells are damaged or nonfunctioning, they must be
   replaced, but generation of new cells must be compensated by
   cell death to maintain a stable baseline population. Such homeostasis is required to maintain cell number and normal function. If
   the equilibrium is disturbed, then abnormal growth and tumors or
   abnormal cell loss can result. A complex system of controls tightly
   regulates homeostasis. One signaling mechanism that operates in
   this regard is the Shh-signaling pathway that normally sends an
   antiapoptotic signal to allow for cell survival. Failure to receive the
   signal results in apoptosis. However, when the hedgehog system
   is impaired, the antiapoptotic signal can be sent inappropriately,
   allowing damaged cells to escape death, with the potential for
   development of malignancy.

B. Initiation of apoptosis

Specific details of apoptotic mechanisms are cell-type and stimulus
dependent; however, research suggests that there are common steps
in this process. Both internal and external death programs exist and
then both use the same downstream mediators to complete the process of apoptosis.

1. **Apoptosome:** An **internal cell death program** will be initiated if
   irreparable damage has been sustained to cellular components or
   to DNA (Figure 23.7). **Bax**, a proapoptotic protein member of the
   Bcl-2 family, is induced and inserted into the mitochondrial membrane to form a channel to allow cytochrome *c* to exit mitochondria.

   Cytochrome *c* in the cytoplasm triggers the formation of the **apoptosome,** a large protein complex that also requires ATP for its formation. The apoptosome is characteristic of apoptosis triggered
   by internal signals. To form this complex, cytoplasmic cytochrome
   c activates the **apoptotic protease activating factor** (Apaf-1)
   adaptor protein that in turn activates caspase 9. Active caspase
   9 initiates the **caspase** proteolytic cascade that will cleave and
   destroy cellular proteins and DNA in order to cause cell death by
   apoptosis (Figure 23.8).

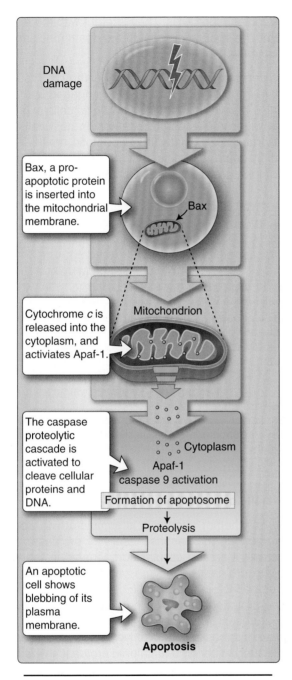

Figure 23.7
Cellular apoptosis via formation of the apoptosome.

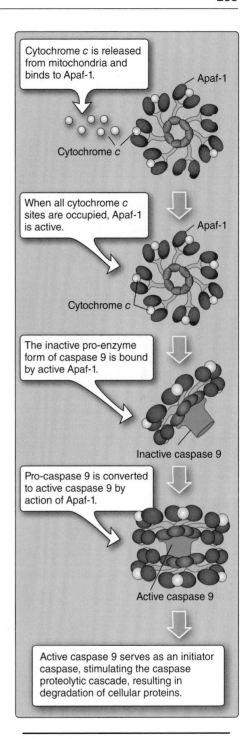

Figure 23.8
Cytochrome *c* activation of Apaf-1 and apoptosome formation.

2. **Death receptors:** The **external program** that stimulates apoptosis occurs via **death receptors** that are members of the **tumor necrosis factor receptor** (TNFR) gene superfamily. Individual members of this family recognize specific ligands, but not all members of the TNF family initiate cell death. Those that do initiate cell death possess a homologous cytoplasmic sequence termed the "**death domain (DD)**." Adaptor molecules such as FADD (Fas-associated death domain) and TRADD (TNFR-associated

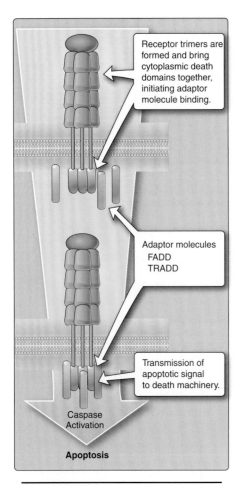

Figure 23.9
Death-receptor initiation of apoptosis.

protein) contain such DDs. They interact with the death receptors to transmit the apoptotic signal to the death machinery, via activation of caspase 8 or 10 (Figure 23.9).

The **Fas death receptor** is a member of the TNFR superfamily that will initiate apoptotic cell death when engaged by the **Fas ligand** (FasL, also known as CD178). T-cytotoxic cells express FasL that interacts with the Fas death receptor on host cells infected with virus in order to stimulate their apoptotic death. TNFR1 is also involved in death signaling, but its death-inducing capability is weak compared to that of Fas (CD95).

Fas ligand–induced apoptosis

Membrane-anchored FasL trimer on the surface of an adjacent cell causes trimerization of the Fas receptor (Figure 23.10). This results in the clustering of the receptors' DDs, which then recruit the cytosolic adaptor protein FADD by binding to FADD's death domains. FADD contains not only a DD but also a death effector domain (DED) that binds to an analogous domain repeated in tandem within procaspase 8, the inactive or zymogen form of caspase 8. The complex of Fas receptor (trimer), FADD, and caspase 8 is called the death-inducing signaling complex (DISC). Upon recruitment by FADD, procaspase 8 is able to activate itself. Caspase 8 then activates downstream caspases and commits the cell to apoptosis. Apoptosis triggered by FasL-Fas (CD178:CD95) plays a fundamental role in the regulation of the immune system.

C. Caspase family of proteases

Regardless of whether apoptosis is stimulated via apoptosomes or by death receptors, the caspase family of proteases will be stimulated to actually degrade cellular components in the apoptotic cell. Caspases are proteases (enzymes whose substrates are proteins) that are major effectors of apoptotic cell death. They are members of the cysteine protease class, which is named after a cysteine amino acid residue present within the catalytic site of the enzyme molecule. Caspases are synthesized as inactive zymogen or proenzyme forms and are activated to become functional proteases when needed. This posttranslational modification ensures that the enzymes can be activated rapidly when required in an apoptotic cell.

1. **Classification of caspases:** Caspases are grouped based on their function (Figure 23.11). Eleven members of the caspase family have been identified in humans. Some are not involved in apoptosis. Caspase 1 is involved in cytokine maturation, caspases 4 and 5 are involved in inflammation, and caspase 14 is important in skin development. The remaining caspases are involved in apoptosis and are grouped into either the initiator or the effector families of apoptotic caspases.

Initiator caspases include caspases 2, 8, 9, and 10. These possess characteristic regions or domains such as caspase recruitment domains (CARD) in caspases 2 and 9 and DED in caspases

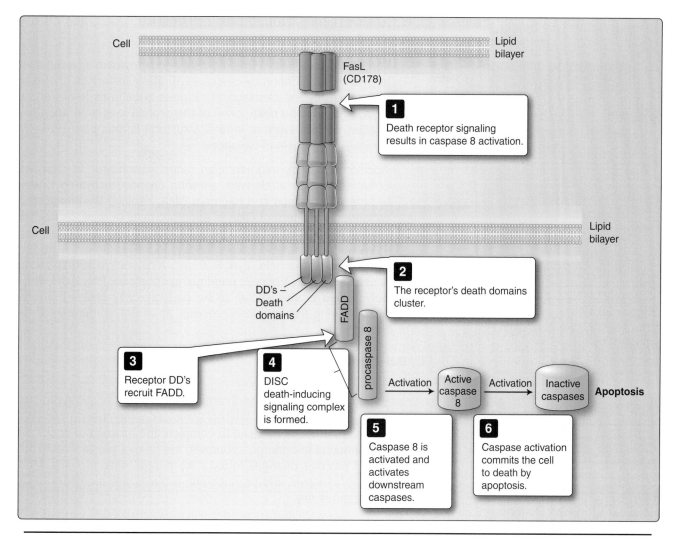

Figure 23.10
Apoptosis as a result of FasL binding to the Fas receptor.

8 and 10 that enable the protease to interact with molecules that regulate their activity. Initiator caspases cleave inactive proenzyme forms of effector caspases, resulting in their activation.

Effector caspases include caspases 3, 6, and 7. These "executioner caspases" proteolytically cleave protein substrates with the cell, causing the apoptotic demise of the cell.

2. **The caspase cascade:** This process is the sequential proteolytic activation of one caspase after another in an orderly fashion during the initiation of apoptosis. Caspase inhibitors regulate the process. The cascade can be activated by various stimuli, including the apoptosome, death receptors, and granzyme B released by cytotoxic T cells.

The apoptosome and death receptors both activate initiator caspases, but different ones. While the apoptosome activates caspase 9, death receptors activate caspases 8 and 10. Granzyme B,

Caspases:	Role:
Caspase 1	Cytokine maturation
Apoptotic:	
Caspases 2,8,9,10	Initiator caspases
Caspases 3,6,7	Effector caspases
Caspases 4,5	Inflammation
Caspase 14	Skin development

Figure 23.11
Classification and roles of caspases.

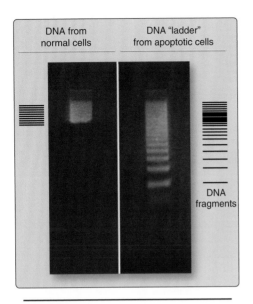

Figure 23.12
DNA fragmentation or laddering in apoptotic cells.

released from cytotoxic T cells, activates caspases 3 and 7, which are effector caspases.

3. **Targets of caspases:** Both nuclear and cytoplasmic proteins are targets of degradation by caspases. In many instances, the exact role played by the cleavage of caspase substrates is not understood and how the destruction of the protein relates to apoptosis is often unclear. Nuclear lamins, structural fibrous proteins in the nucleus, are targets of caspases.

Additionally, DNA fragmentation factor 45/inhibitor of caspase-activated DNAse is cleaved, allowing caspase-activated DNAse to enter the nucleus and fragment DNA, causing the characteristic laddering pattern of DNA in apoptotic cells (Figure 23.12). The DNA is cleaved by an endonuclease into fragments that are multiples of the same size, corresponding to the length of the nucleosome coil, for example, 2, 4, 6, 8, etc. A distinctive 180-bp ladder is seen in the DNA of cells undergoing apoptosis. Poly ADP ribose polymerase is also known to be proteolytically cleaved by caspases during the apoptotic process, as is Bid, a member of the Bcl-2 family.

D. Bcl-2 family

As is also true for caspases, regardless of whether apoptosis has been initiated in a cell by an internal apoptosome-requiring program or via death receptors and external stimulation, members of the Bcl-2 family of proteins will be used to complete the process of apoptosis. In cells induced to undergo apoptosis, the ratio of prosurvival to proapoptotic proteins changes to favor the proapoptotic proteins. Many of these prosurvival and proapoptotic proteins are members of the Bcl-2 protein family.

Prosurvival (antiapoptotic) members of the Bcl-2 family include **Bcl-2** and **Bcl-xL,** while prodeath members include **Bak** and **Bax** (Figure 23.13).

When the internal program of the apoptosome stimulated apoptosis, the prodeath **Bax**, is induced and inserted into the mitochondrial membrane to form a channel to allow cytochrome *c* to exit mitochondria. Death receptor signaling normally results in the activation of caspase 8, which catalyzes the cleavage of Bid to tBid (Figure 23.14). In response, Bak and Bax can then translocate from the cytosol to the outer mitochondrial membrane, permeabilize it, and facilitate the release of proapoptotic proteins including cytochrome *c* and Smac/DIABLO, an antagonist of inhibitors of apoptosis proteins (IAPs).

E. Apoptosis in disease

Apoptosis is required for normal development and physiological function, so necessarily mutations in mediators of apoptosis can result in diseases ranging from cancer (insufficient apoptosis) to Alzheimer disease (excessive apoptosis). Some other neurodegenerative diseases associated with aging including Huntington disease and Parkinson disease involve apoptotic cell death of normal, functional cells whose survival would benefit the individual.

Bcl-2 Family
Pro survival:
Bcl-2
Bcl-xL
Pro death:
Bax
Bak
Bid

Figure 23.13
Prosurvival and prodeath members of the Bcl-2 family.

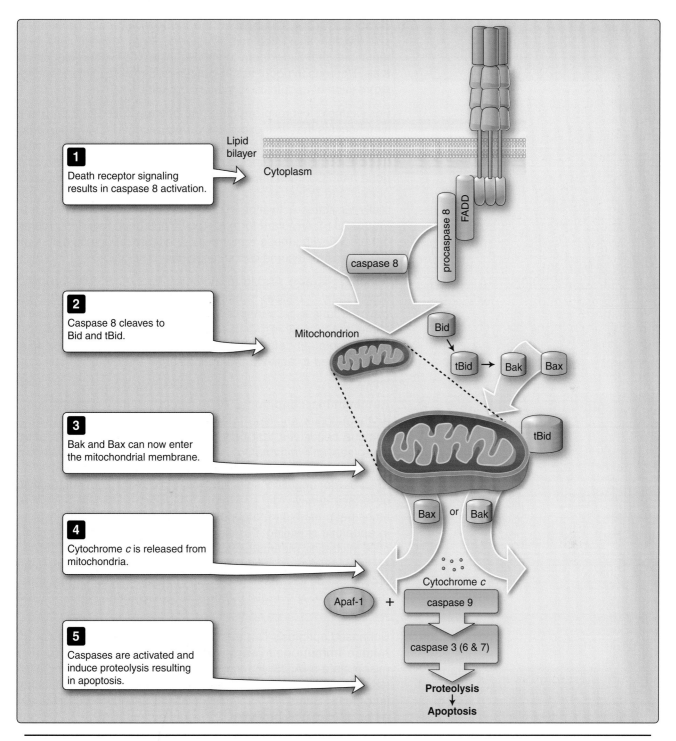

Figure 23.14
Bcl-2 family members in apoptosis.

1. **Cancer:** Cancer arises when the homeostatic balance between cell division and apoptosis is altered. Many cancer cells have mutations that allow them to survive instead of undergoing apoptosis. Decreased apoptosis in cancer cells may result from altered expression of apoptosis-regulatory proteins.

   For example, overexpression of the antiapoptotic Bcl-2 protein in lymphocytes that also express the *myc* oncogene and when chromosomal translocations move the Bcl-2 gene next to the immunoglobulin heavy chain locus, lymphoma can result. The Bcl-2 gene has also been implicated in breast, prostate, and lung carcinomas and in melanoma. Another example is Apaf-1, which normally facilitates activation of caspase 9 by the apoptosome (see Figure 23.8). A mutant form of Apaf-1 has been found in prostate tumors, allowing these tumor cells to escape apoptotic cell death since the apoptosome cannot assemble properly.

   Apoptosis is also related to therapies for cancer. Ionizing radiation therapy is believe to help reactivate dormant apoptotic pathways of cancer cells. And resistance to cancer chemotherapy in some tumors may also be caused by the overexpression of Bcl-2 and subsequent defective apoptosis. Some new cancer therapies are designed to activate the apoptosome and others to stimulate caspases.

2. **Autoimmune conditions:** Rheumatoid arthritis, systemic lupus erythematosus, and type 1 diabetes mellitus are examples of autoimmune conditions that may result in part from defective apoptosis (see *LIR Immunology*, Chapter 16). A prominent feature of autoimmune diseases is the failure of T cells that recognize self to undergo negative selection via apoptosis. The autoreactive cells then survive and proliferate. Defective apoptosis has sometimes been attributed to abnormal expression of proteins involved in apoptotic signaling. For example, some studies indicate that T cells infiltrating rheumatoid synovium express high levels of Bcl-2 and are resistant to Fas-induced apoptosis. Defective Fas death receptors and increased apoptosis in pancreatic islets may cause destruction of pancreatic β cells and development of type 1 diabetes mellitus.

   Enhanced apoptosis can also impact progression of infection with **human immunodeficiency virus** (HIV) to the immune-compromised state of AIDS. Inappropriate apoptotic depletion of CD4$^+$ T helper cells causes marked decreases in these T cells in affected individuals. Some HIV proteins inactivate antiapoptotic Bcl-2, and other HIV proteins promote Fas-mediated apoptosis.

3. **Neurodegenerative illnesses:** Other types of disorders may also result in part from increased apoptosis. **Schizophrenia**, a chronic neurodegenerative illness, is characterized by delusions, hallucinations, and changes in emotional state. Although the mechanisms underlying these deficits are largely unknown, recent postmortem data implicate a role for altered neuronal apoptosis. Apoptotic regulatory proteins and DNA fragmentation patterns appear to be altered in several cortical regions in individuals with schizophrenia. In individuals with **Alzheimer disease**, localized apoptosis may

contribute to early neurite and synapse loss, leading to the initial cognitive decline. Additionally, many individuals infected with the HIV virus develop a syndrome of neurologic deterioration known as **HIV-associated dementia** (HAD). HAD appears to be associated with active caspase 3 in the affected brain regions, leading to speculation that pharmacologic interventions aimed to inhibit the caspase pathway may be beneficial in halting the destructive apoptosis in these regions.

F. Laboratory assessment of apoptosis

Several laboratory assays exist to assess apoptosis. In addition to the methods described below, analysis of expression of proapoptotic proteins, such as Bax, and measurement of caspase activity can also be done.

1. **DNA laddering:** Visualization of **DNA laddering** is perhaps the oldest technique available to detect that apoptosis has occurred. Since the genomic DNA of apoptotic cells is degraded into approximately 180 base pair fragments, a characteristic laddering appearance is revealed on agarose gel electrophoresis (see Figure 23.11).

2. **TUNEL:** The **TUNEL** method, which stands for *t*erminal *u*ridine deoxy*nu*cleotidyl transferase nick *e*nd *l*abeling, detects DNA fragmentation based on the presence of strand breaks or nicks in the DNA. The terminal deoxynucleotidyl transferase enzyme catalyzes the addition of dUTPs that have been labeled for the experiment.

3. **Annexin 5:** The **Annexin 5** affinity assay is also useful for the detection of cells early in the apoptotic process. Annexins are a family of proteins that bind to phospholipids in cell membranes. Annexin 5 binds to phosphatidylserine, which, in healthy cells, is present on the inner membrane leaflet. Soon after a cell has initiated steps toward programmed cell death, phosphatidylserine flip-flops to the outer membrane leaflet. A labeled antibody to Annexin 5 can be used to detect cells displaying phosphatidylserine on their outer leaflet, indicating that they have initiated the apoptotic process.

4. **Flow cytometry:** This procedure can be used to measure **cell size** and **granularity** of cells within a population, both of which differ in apoptotic and normal cells. Because apoptotic cells shrink in size, the forward angle light scatter will reveal an apoptotic population of less intensity compared with normal cells. Granularity of apoptotic cells is increased compared with that of normal cells, as indicated by side scatter.

Chapter Summary

- Cell death occurs via one of the two processes: necrosis, a pathological process, or apoptosis, a physiological process.
- Necrotic cells rupture and release their contents, including enzymes, to the extracellular media, often inducing an inflammatory response.
- Apoptotic cells undergo a programmed form of cell death with blebbing membranes that remain intact. Their engulfment by phagocytic cells prevents their death from causing an inflammatory response.
- Apoptosis is important for elimination of damaged cells in development and for tissue homeostasis.
- Apoptosis may be stimulated by an internal process resulting in assembly of an apoptosome or by extracellular signals via death receptors.
- Apoptotic cells show a change in their ratio of proapoptotic and prosurvival protein members of the Bcl-2 protein family to favor the proapoptotic proteins.
- Caspase proteases catalyze cleavage of cellular proteins, culminating in apoptosis.
- Insufficient apoptosis in certain cells can result in cancer and autoimmune conditions, while excess apoptosis can lead to neurodegenerative illnesses and may play a role in the development of AIDS from HIV infection.

Study Questions

Choose the ONE best answer.

23.1 A 64-year-old male is suspected of having experienced a massive lysis of his erythrocytes. Positive results for which of the following tests may help confirm this suspicion?
   A. Annexin 5
   B. Caspase activity
   C. DNA laddering
   D. LDH in blood serum
   E. TUNEL assay

Correct answer = D. Massive lysis of erythrocytes occurs by necrosis, which is accompanied by the release of intracellular enzymes, including LDH, into the patient's blood. Assays of annexin 5, caspase activity, DNA laddering, and TUNEL are all used to detect apoptotic cells. Since erythrocytes do not contain genomic DNA, measurement of DNA laddering of erythrocytes is never possible.

23.2 An 8-year-old female sustained an injury to her arm, which initiated an inflammatory response, resulting in pain and swelling. Which of the following is characteristic of the cells that died to initiate this response?
   A. Activation of caspase 3 to cleave cellular proteins
   B. Cytochrome c release from mitochondria
   C. Fas death receptor binding to an extracellular ligand
   D. Increase in the intracellular ratio of Bax to Bcl-2
   E. Plasma membranes that have ruptured during the process

Correct answer = E. Ruptured plasma membranes are characteristic of necrotic cells that induce an inflammatory response as a result of their death. The other answer choices all relate to apoptotic cell death that does not stimulate inflammation.

23.3 A cell has been induced to die in the course of normal development, and an apoptosome forms within that cell. Which of the following is required for this process?
   A. Energy in the form of ATP
   B. Fas death receptor signaling
   C. Loss of mitochondria from cells
   D. Plasma membrane rupture
   E. Release of enzymes from cells

Correct answer = A. Apoptosome formation in response to internal apoptotic signals requires energy in the form of ATP. FasL signals via Fas death receptors when apoptotic signals are received from external sources. Loss of mitochondria, plasma membrane rupture, and release of enzymes all occur during necrotic cell death.

23.4 A population of cells is being grown in the laboratory, and the total number of viable cells is observed to decrease. Analysis of Bax protein reveals an increase in its expression. Which of the following findings may be used to confirm that apoptosis is occurring in this cell population?
A. Increased expression of Bcl-2 protein
B. Intracellular enzymes in the culture medium
C. Phosphatidylserine on the outer membrane leaflet
D. Release of mitochondria from the dying cells
E. Visualization of ruptured plasma membranes

Correct answer = C. Phosphatidylserine is a phospholipid normally present on the inner leaflet of a plasma membrane. But, during apoptosis, it appears on the outer leaflet. Annexin 5 binding to phosphatidylserine can also be used to detect it on the outer membrane. Bcl-2 is antiapoptotic and would not increase its expression during apoptosis. Necrotic cells whose plasma membranes rupture release the intracellular enzymes and mitochondria.

23.5 In which of the following conditions may high levels of apoptotic cell death result in disease?
A. Breast cancer
B. HIV/AIDS
C. Lymphoma
D. Rheumatoid arthritis
E. Systemic lupus erythematosus

Correct answer = B. Progression of HIV infection to AIDS involves apoptotic depletion of CD4$^+$ helper T cells. Insufficient apoptosis is observed in cancers, such as breast cancer and lymphoma, and also in autoimmune diseases, such as rheumatoid arthritis and systemic lupus erythematosus.

24 Aging and Senescence

I. OVERVIEW

Aging is reflected by the changes that occur over the lifespan of an individual. It is also described as a decline in biological function over time. There are several hypotheses to explain the aging process in eukaryotic cells. Characteristics of aging, including graying hair, wrinkling skin, and diminishing eyesight, affect all individuals who reach more advanced age and are separate from diseases that result due to the aging process, such as heart disease, type II diabetes mellitus, and cancer.

Senescence is the process by which normal, diploid eukaryotic cells lose the ability to divide, contributing to the process of aging. Aging and senescence are related, as cells from adults can divide fewer times than cells from neonates. Our cells do not retain the ability to divide indefinitely but, after a certain number of divisions, depending on the cell type, they will no longer be able to reproduce themselves. After this has happened, cells are described as being in a state of replicative senescence. Several genes are known to play a role in the control of replicative senescence. While senescence of certain types of cells in younger individuals may protect from the development of cancer, over time, when enough senescent cells have accumulated, phenotypes of aging and their associated pathologies result.

II. ENTERING INTO SENESCENCE

Senescent cells are alive, functioning, and metabolically active, but they can no longer divide. The finite number of divisions human cells can undergo was first described by Dr. Leonard Hayflick in the 1960s using data from human fibroblasts. His findings revealed that replicative senescence depends on the number of cell divisions that the cell has completed during its lifetime—not on the total time it has spent in active phases of the cell cycle. It is also dependent on the cell type. While some cells, such as the germ line cells, divide indefinitely, most other normal cells in the body will stop dividing and enter into a nondividing senescent stage as they age.

Proliferating cells reach their limit in number of cell divisions that they can undergo largely because, after repeated cycles of DNA replication, they lose the ability to duplicate DNA all the way to the end of the entire chromosome, causing telomeres to progressively shorten (see also Chapter 7). Cells with shortened telomeres then normally enter into senescence.

A. Failure to become senescent

Telomerase reverse transcriptase is part of the telomerase enzyme complex and acts to lengthen telomeres by adding repetitive nucleotides in a TTAGGG sequence at the end of the telomere, protecting it from degrading after multiple rounds of cell division. If this enzyme acts in cells that should have entered senescence, then they may continue to divide.

Cells with shortened telomeres normally enter into senescence. But those that fail to become senescent and continue to proliferate will often develop aberrations in their chromosomes that can result in cancer. Therefore, the senescence response represents a fail-safe mechanism in place to help to prevent malignant transformation. Accumulation of DNA damage, inappropriate expression of oncogenes, and generation of reactive oxygen species (ROS) are all known to induce senescence (Figure 24.1).

B. The senescent phenotype

Senescent cells display a number of characteristics that distinguish them from quiescent resting cells that are still able to divide. Senescent cells are considered to be in a terminally differentiated state that cannot be reversed by physiological stimulators of growth. Senescent cells, while incapable of replication, are generally viable and metabolically active, survive long term, and resist apoptosis (Figure 24.2). These cells also express a senescence-associated beta-galactosidase enzyme activity (SA-B-gal).

1. **Irreversible cell cycle arrest:** Growth arrest of senescent cells occurs mostly in the G_1 phase and is accompanied by an increased expression of cell cycle inhibitors and a decreased expression of cell cycle regulators such as cyclins and transcription factors (E2F) (see Chapter 21) that are necessary for normal progression through the cell cycle.

   The cyclin-dependent kinase inhibitors p21 and p16 are important mediators in aging and age-related disease. They function to inactivate the cyclin-dependent kinases (Chapter 21), which maintain the retinoblastoma protein (pRB) in its active hypophosphorylated state necessary to block cell cycle progression in G_1.

2. **Chromatin modification: Chromatin remodeling** is modification of chromatin structure to regulate the ability of transcription machinery to bind to DNA, and is a means to control gene expression. Deficiencies in chromatin remodeling occur with aging. Senescent cells are further characterized by:

   a. **Changes in DNA methylation:** Overall, there is a genome-wide decrease in DNA methylation with age, specifically in regions of repetitive DNA sequences. However, there is also an increase in methylation at CG sites found within promoter regions of genes.

   b. **Deacetylation of histones:** This type of change to histones creates unique regions in the chromatin called **senescence-associated heterochromatin foci** (SAHF). SAHF participate

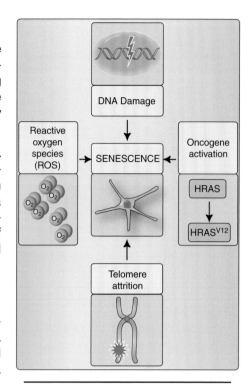

Figure 24.1
Senescence inducers.

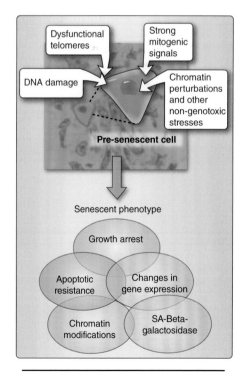

Figure 24.2
The senescent phenotype.

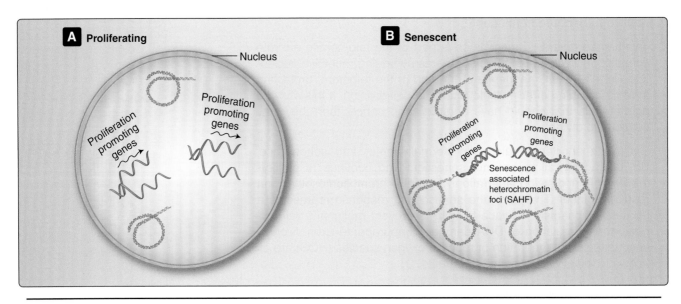

Figure 24.3
Senescence-associated proliferation arrest.

in senescence-associated proliferation arrest by sequestering proliferation-promoting genes, including cyclin A, which is needed for cells to progress through S phase. SAHF are not associated with reversible cell cycle arrest in quiescent cells (Figure 24.3).

3. **Sirtuins: Sirtuins** are a family of proteins that possess NAD^+-dependent deacetylase activity and participate in regulating cellular metabolism and in protecting DNA from age-related changes. Sirtuins are evolutionarily conserved, found in roundworms, fruit flies, yeast, and mammals. The first sirtuin discovered, silent information regulator 2 (Sir2), was identified in yeast as a deacetylase. Since then, other sirtuins have been found to have additional types of enzymatic activities, including lipoamidase and ADP-ribosyltransferase activities, which also require NAD^+. Because of their dependency for NAD^+, they are believed to be important regulators in cellular energy metabolism.

It appears the sirtuins have evolved to respond to the availability of NAD^+, which, in addition to playing an important role in energy metabolism, is essential for repair of DNA damage. When NAD^+ declines in cells, sirtuin activity decreases, a phenomenon seen in aging. Systemically, when NAD^+ biosynthesis decreases with age, it appears that communication between the hypothalamus (aging control center) and adipose tissue (its modulator) is disrupted.

Recent reports have verified previous speculation that sirtuins regulate aging and longevity. For example, when the brain and hypothalamus of mice are made to overexpress Sirtuin 1 (SIRT1), the human counterpart of Sir2, the mice show delayed signs of aging and increased lifespan. And mice made to overexpress another sirtuin member, SIRT6, have been shown to have longer lifespans.

4. **Senescence-associated secretory phenotype (SASP):** Senescent cells remain metabolically active and undergo several changes in gene expression, resulting in the production of specific proteins that are secreted by these cells. These include proinflammatory cytokines, growth factors, chemokines, and extracellular matrix–degrading enzymes that are collectively referred to as the senescence-associated secretory phenotype, SASP. The components may vary according to cell types, but the consequence is tissue deterioration of extracellular matrix and inflammation. Chronic inflammation and tissue dysfunction as a result of SASP appear to drive aging and its associated chronic conditions such as type II diabetes mellitus, neurodegenerative disease, and cancer.

Clinical Application 24.1: Calorie Restriction, Aging, and Red Wine

Caloric restriction, which is the reduction of caloric intake without malnutrition, is known to be important for extending the lifespan in a wide range of species including humans. Consumption of fewer calories may also help postpone manifestations of aging, including both functional decline and age-related diseases. Much is known about the physiological changes that occur when individuals restrict the number of calories they consume, but much less is understood about the molecular mechanisms involved. Sirtuins may be key molecules that affect longevity within the context of calorie restriction; sirtuins have been found to be induced by calorie reduction. Resveratrol, a polyphenolic compound in red wine, which has been shown to mimic effects observed in calorie restriction, potently induces SIRT1 expression. In response, a number of related protective effects on cell metabolism occur that assist in inhibiting age-related decline in physiological functions. The amount of resveratrol necessary is the caveat—an equivalent of 100 bottles of red wine is required to produce the beneficial effect seen in laboratory animals!

Clinical Application 24.2: Progerias and Nucleus Architecture

Hutchinson-Gilford progeria syndrome (HGPS), a rare syndrome caused by a mutation in the Lamin A gene that encodes a nuclear scaffold protein, is characterized by greatly accelerated aging. Affected individuals demonstrate severe growth retardation, loss of subcutaneous fat, reduced bone mineral density, alopecia, and poor muscle development. They also show wrinkled skin and have a higher incidence of stroke and myocardial infarction. The average life expectancy for individuals with this syndrome is 12 to 15 years. Interestingly, these patients do not present with all aspects of aging. In fact, they do not show increased incidence of cancers, neurodegeneration, or other age-related disorders such as arthritis or cataracts. But, at the molecular level, accelerated aging is noted in cells of patients with HGPS that show dramatic changes in structure (Figure 24.4) and in chromatin that are similar to those seen in the healthy, but much older persons.

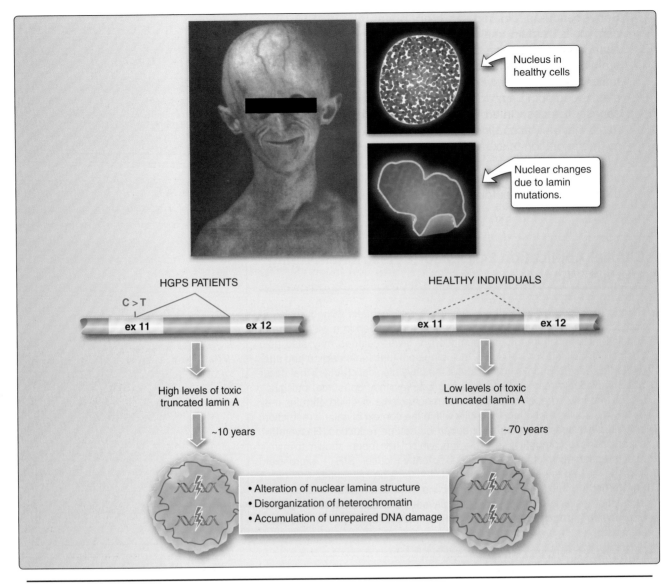

Figure 24.4
Hutchinson-Gilford progeria syndrome (HGPS) and nuclear architecture.

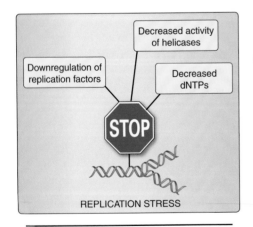

Figure 24.5
DNA replication stress.

C. Mechanisms involved in aging

Several theories have been proposed to account for changes that occur as humans age. However, the specific contribution of each and their relative importance to the overall aging process continues to be a matter of debate.

1. **DNA replication stress:** Defined as inefficient DNA replication, DNA replication stress causes DNA replication forks to progress slowly or to stall. Many factors are known to contribute to this type of stress to DNA replication, including down-regulation of replication factors, decreased dNTPs, and decreased activity of helicases (Figure 24.5).

RecQ helicases are important in maintaining the genome and function in unwinding DNA and aiding in replication of double-stranded DNA. With decreased helicase activity, replication does not proceed normally and contributes to the process of aging. Mutations in the WRN RecQ helicase gene cause the premature aging syndrome **Werner syndrome**. In addition to shortened lifespan, individuals with Werner syndrome exhibit premature graying and thinning of hair, osteoporosis, type II diabetes, cataracts, and an increased incidence of cancer.

2. **Mitochondria, reactive oxygen species, and aging:** Mitochondria have their own genome, and they replicate and transcribe their own DNA, independent of the nucleus. Like nuclear DNA, mitochondrial DNA (mtDNA) is constantly exposed to DNA-damaging agents. The free radical theory of aging proposes that aging and the associated degenerative diseases could be attributed to deleterious effects of free radicals on cellular components. One of the sources for ROS in the cell are products of oxidative phosphorylation that occurs within mitochondria. Therefore, the free-radical theory of aging is essentially a mitochondrial theory of aging. Because mtDNA is vulnerable to oxidative stress due to the close proximity between the sites of ROS production and mtDNA, mutations in mtDNA accumulate progressively during life. Such mutations are directly responsible for a measurable deficiency in cellular oxidative phosphorylation activity, leading to an enhanced ROS production. The increase in mtDNA DNA with age, combined with increased ROS production, creates a "vicious cycle" that culminates in the aging of the organism (Figure 24.6).

3. **Stem cell theory of aging:** According to the stem cell theory of aging, tissue-specific adult stem cells, which are important for self-renewal and tissue replacement during the human lifespan, show an age-associated decline in the ability to divide. These cells are generally retained in the quiescent state but can be induced into the cell cycle in response to physiological growth factor stimulation even after prolonged periods of dormancy. Once stimulated, the stem cells can produce undifferentiated progeny, which, in turn, produce differentiated cells through subsequent rounds of proliferation. It is thought that aging affects the ability of stem cells to produce both the undifferentiated progeny and the differentiated cells (Figure 24.7).

D. Molecular mechanisms in the senescence response

While diverse stimuli can induce the senescence response, they all appear to converge on either one or two different pathways that both establish and maintain the senescence response. These pathways are governed by two tumor suppressor proteins, p53 and pRB (Chapter 21). P53 activates differentiation in the self-renewing tissue-specific stem cells. The tumor suppressor P16 has also been shown to play an important role in the age-related decline of stem cells; an age-associated increase in p16 levels restricts self-renewal and upsets tissue homeostasis.

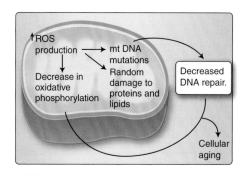

Figure 24.6
Mitochondria, reactive oxygen species, and aging.

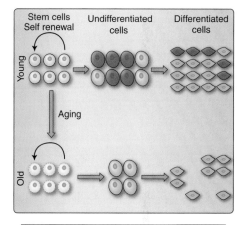

Figure 24.7
Stem cell's functionality is altered with age.

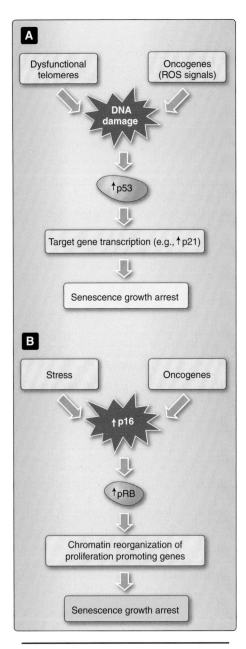

Figure 24.8
Molecular mechanisms in senescence response.

1. **The p53 pathway:** It is well known that p53 is an important mediator of cellular responses to DNA damage. P53 is an important mediator of the senescence response since attrition of telomeres, which resembles damaged DNA, also triggers an increase in p53. Inappropriate activation of oncogenes within normal cells also induces a senescence response through activation of p53. Oncogenes such as Ras (Chapters 17 and 22) signal through generation of ROS, which is necessary for their mitogenic effects. However, generation of DNA-damaging ROS also activates the p53-dependent damage response.

   At least in some cell types, the induction of senescence by DNA damage, telomere dysfunction, and possibly oncogene overexpression converge on the p53 pathway, which is both necessary and sufficient to establish and maintain senescence arrest (Figure 24.8A). Although the senescence-induced growth arrest cannot be reversed by physiological growth signals, it can be reversed by inactivation of p53, which explains, in part, the age-related increase in cancer.

2. **The pRB pathway:** In some cells, p53 inactivation alone appears to be insufficient to reverse the senescent phenotype. The difference appears to be in the presence or absence of p16 expression. Stress increases the expression of p16, and the resultant increase in pRB is known to aid in the reorganization of the chromatin, resulting in senescence-related inhibition of gene encoding cell cycle regulators (Figure 24.8B).

Chapter Summary

- Most eukaryotic cells undergo a finite number of cell divisions during their lifespan.
- Replicative senescence refers to a permanent halt in the ability to proliferate and is determined by the number of cell divisions that a cell has completed.
- Senescent cells are metabolically viable but will not enter active phases of the cell cycle under any condition.

Chapter Summary (Continued)

- Senescent cells show increased expression of cell cycle inhibitors such as p16 and p21.
- The chromatin in senescent cells demonstrates a unique structure as a result of modifications to histones and DNA.
- Sirtuins are a group of proteins with NAD-dependent deacetylase activity that offer protection against age-associated declines in metabolism and physiological functions.
- Mitochondria-associated ROS formation is implicated in the oxidative damage to macromolecules coupled with a decrease in oxidative phosphorylation.
- Age-associated decline in the replicative potential of stem cells also accounts for the deficiencies in tissue replacement during aging.

Study Questions

Choose the ONE best answer.

24.1 Senescent cells

    A. Undergo apoptosis instead of dividing.
    B. Experience irreversible cell cycle arrest.
    C. Retain full-size, long telomeres
    D. Are metabolically inactive.
    E. Have undetectable beta-galactosidase activity.

Correct answer = B. Senescent cells experience irreversible cell cycle arrest. They show telomere attrition (shortened telomeres), resist apoptosis, express beta galactosidase, and remain metabolically viable. They are, however, unable to enter the cell cycle under any condition.

24.2 Which of the following types of DNA modifications are known to be associated with aging?

    A. Cytosine methylation of DNA in promoter regions
    B. Deacytelation of chromation regions
    C. Oxidative damage to mtDNA
    D. Increased DNA methylation at CG sites
    E. All of the above

Correct answer = E. All of the above. Methylation of cytosines in promoter regions in DNA is important to silence the critical genes, while deacetylation of chromatin regions creating the SAHF results in the silencing of a number of cell cycle regulatory genes. Oxidative damage due to generation of ROS and subsequent mutations in mtDNA are implicated in the aging process. Methylation and demethylation of DNA are seen with aging, but methylation appears to be limited to the repeat CG sequences in DNA.

24.3 An increase in p53 in adult stem cells is associated with

    A. Activation of proliferation.
    B. Apoptosis.
    C. Generation of ROS.
    D. Induction of differentiation.
    E. Shortening of telomeres.

Correct answer = D. An increase in p53 in adult stem cells is associated with an induction in differentiation of these cells, not proliferation or apoptosis. Increased p53 does not cause a generation of reactive oxygen species (ROS) or a shortening of telomeres.

Index

Note: Page numbers followed by *f* indicate figure; those followed by *t* indicate table.